I0797570

OVER THE WALL

OVER THE WALL

A BRIXMIS INTELLIGENCE OFFICER BEHIND THE IRON CURTAIN

WILL BRITTEN

Map of the former DDR.

Jacket illustrations: Front: The Berlin Wall at Potsdamer Platz in 1979. (2ebill/Alamy Stock Photo) *Back:* The author with a trio of fraternal comrades. (Author photo)

First published 2025
The History Press
97 St George's Place, Cheltenham,
Gloucestershire, GL50 3QB
www.thehistorypress.co.uk

British Library Cataloguing in Publication Data.
A catalogue record for this book is available from the British Library.

ISBN 978 1 80399 856 5

Typesetting and origination by The History Press
Printed and bound in Great Britain by TJ Books Limited, Padstow, Cornwall

Trees for Life

To those who watched and to those we watched.
To Roy P, who inspired me to join the unit, and to
Mark Lucas for sowing this literary seed.
And to my children.
(*May 2021*)

The names of all Allied military personnel have been changed to protect their privacy and, in some cases, their personal security, unless they are already in the public domain. Some of these characters appear in *The Deadly Game*. I have used the names of Soviet officers as they were given to us.

'In the fields of observation chance favours only those minds which are prepared.'
Louis Pasteur – presciently capturing
the essence of Military Mission 'touring'.

Contents

Preface

The crack and thump of three high-velocity shots hammered against the warm, taut air of the early spring afternoon.

It was 24 March, a Sunday, two years after the fear and tension that scarred 1983, and four and a half years before the disbelief and ecstasy of the 9 November revolution – still the far-off dream that no one yet had had the courage to conjure with – when the Berlin Wall came tumbling down.

The Soviet command, sitting in Zossen Wünsdorf, just 50km south of Berlin, had imposed a temporary restricted area (TRA) to cover a large-scale exercise that the Allied Military Missions' operations staffs assessed could involve the assets of up to three separate German Democratic Republic-based Red Army divisions. A two-man tour had crossed the Glienicke Bridge in the early hours of the morning from the light and freedom of West Berlin into the pervading greyness of Potsdam, timed to hit the southern fringes of the exercise area as the TRA was scheduled to lift. Their tasking was to move northwards, to rummage through pre-selected areas and specific locations in the hopes of finding documentary or other intelligence left behind by the units as they withdrew from the exercise area and returned to barracks.

A quick check of this sector of the training area by the tour crew had discovered signs of activity but nothing of intelligence value worth recovering back to West Berlin. The tour officer in command paused a moment, once back in the safety and security of his military green Mercedes G-Wagen, to decide on the team's next move before heading north to remain in sync with the tour plan, the intent to visit Neubukow, near the port of Rostock, by dark before completing their 1,500km sweep and returning back to base late the following evening. He had just three or four more tours at most before his posting in the Mission was over and he moved on to his next assignment – he was keen to maximise the intelligence he brought home before his posting. One more push and maybe one more success.

He decided they would call in quickly at Ludwigslust, 190km into the heartland of the Deutsche Demokratische Republik (DDR), from where the team had crossed over from West Berlin into the east. Ludwigslust was home to a Motor Rifle Regiment subordinated to 21 Motor Rifle Division, a formation that had often in the past been one of the first units in the Group of Soviet Forces in Germany (GSFG) to receive new equipment entering front-line service in the DDR. It was to be one of the last decisions he made as a Mission tourer and the most fateful of his life.

Just sitting on the edge of a permanently restricted area (PRA) were two small training areas. One of them had a small shed capable of temporarily housing up to eight tanks or self-propelled tracked artillery guns. On a Sunday it was possible that it could be full, awaiting a resumption of normal training the following morning. Even hard-pressed Red Army conscripts had one day of rest per week.

The team approached cautiously, as the previous summer a French tour had been shot at whilst surreptitiously checking a tarpaulin-covered vehicle parked outside the same shed. Confirming the immediate area was clear, the tour officer lowered his binoculars and instructed the driver/tour NCO, a staff sergeant, to pull forward slowly, circling the shed to ensure it was clear of sentries on the other, blind, side and to park up so he could slip out and stealthily check the padlocked main door. Neither soldier had spotted the Soviet guard resting out of sight in the shaded treeline, enjoying the spring warmth. But, critically, he had spotted and recognised them and understood their role and mission.

The time was 1.30 p.m. The G-Wagen halted and the officer cautiously alighted from the front passenger seat and pushed shut his door. The driver swung open the turret hatch above him and raised himself up and out to increase his visibility as he watched his commander cover the 40m to the shed door.

He felt the first shot pass close over his head like an angry hornet. Instantly, instinctively, he dropped back onto his seat, explosively pulling closed and securing the hatch, now above his head, in one fluid movement.

Two further shots rang out and there, lying on the ground, face down, was Major Arthur 'Nick' Nicholson, the United States Military Liaison Mission (USMLM) senior ground touring intelligence officer. He raised his head.

'Jesse, I've been shot,' were his last words as he slumped back into the pool of his own blood spreading steadily over the dampened earth.

Staff Sergeant Jesse Schatz, one of the most experienced drivers and tour NCOs in the American Mission, was firmly ordered, at gunpoint, by the Soviet soldiers and young officer, who were soon on-scene taking control of the incident, to remain in the vehicle, despite him holding up the tour's first aid box with its universal red cross stencilled on the lid.

Major Nicholson died of his wounds a short time later, essentially bleeding to death for lack of an effective medical response, as Schatz looked on helplessly, powerless to do more than offer a silent prayer.

This incident could have happened to any Allied tour operating in this northern zone of Mission activity. In the spirit of 'all for one and one for all', the fact it was an American musketeer who tragically lost his life, and not one from either the French Military Liaison Mission or the British Mission (BRIXMIS) was mere chance, a coincidental dance to the music of time. Nick Nicholson had been conducting himself professionally; neither he nor his tour vehicle were in an area where they should not have been, and both his and the tour NCO's conduct posed no direct threat, or the implication of threat, to any member of the nearby Soviet garrison. It was a classic case of wrong place, wrong time, compounded by wrong sentry, wrongly or, more appropriately, poorly trained and briefed. It is even feasible that Russian wasn't even the sentry's first language

But what the incident, ultimately, graphically demonstrated was the inherent danger of conducting reconnaissance missions 'over the Wall', behind the front line, deep inside enemy territory, and the existential risks the Allied Mission tours had faced every single day over the four decades during which they conducted this vital, and unique, intelligence collection operation against the Warsaw Pact, the length and breadth of communist East Germany, throughout Cold War history.

1

Questions (and Answers)

To make some sense of the tragic tale of Major Nicholson's death, and to colour the landscape and create some context, perspective and relief, there are some key assumptions – acknowledgement of the gist of each – which underpin the detail of the 'what, where, when and how' of the historical events leading to that fateful afternoon.

If I pressed you that on every day for forty-four years, from late 1946 to the end of 1990, on-duty members of the British Army, the Royal Air Force and occasional members of the Royal Navy had been engaged on officially sanctioned military operations, travelling throughout East Germany – the DDR – demonstrably *behind* the Iron Curtain, would you believe me?

If I confirmed to you that these British servicemen were also joined by members of the US and French armed forces, would you believe this too?

If I was to persuade you that they travelled overtly in the uniforms of their respective services, in plainly marked civilian vehicles, with, in the British case, Union flags unequivocally visible, does that sound credible?

What about if I continued to tell you their primary *raison d'être* was to systematically conduct acts of espionage – crudely put, to 'spy' – on the Red Army and their comrades-in-arms in the DDR's *Nationale Volksarmee*, clandestinely seeking to report on the operational doctrine and capability of the five massive enemy Soviet ground armies – as they occupied over 750 barracks spread over 270 separate locations, exercising on nearly 120 training areas – and 11 divisions of the NVA based in East Germany? Further, that they routinely conducted intelligence collection missions against these Warsaw Pact barracks, the seemingly never-ending cycle of training exercises, the four dozen airfields from which the Soviet Air Force flew, and all the road and rail movement of equipment, seeking to remain updated on their organisation, tactics and equipment … feasible and credible?

If I put it to you that these teams were regularly under hostile surveillance mounted by the East German Ministry of State Security (MfS) – *Stasi* – frequently shot at by Soviet sentries and guards, and others were assaulted and beaten, and that not only Major Nicholson but another member of the Allied Missions lost their life to enemy action, could you believe that to be true?

And finally, if I stated that teams comprising these soldiers, sailors and airmen were on hand to monitor the Soviet force's reaction to every escalation of tension during the Cold War and to inform their own governments' decision-making process, when, with every turn of the screw, the prospect of war seemingly loomed larger, from the Berlin Airlift to the Cuban Missile Crisis, the invasions of Hungary and Czechoslovakia, the building of the Berlin Wall, the tensions of the Reagan years and, finally, the collapse of the Wall and the overthrow of Gorbachev, would you believe this?

Incredibly, each and every one of these tantalising scenarios is true. They relate to, and shape the story of, the Allied Military Missions – unique and remarkable military intelligence units that ranged across East Germany throughout the duration of the Cold War.

This is essentially, but not exclusively, the story of the British servicemen who belonged to the British Commanders'-in-Chief Mission to the Soviet Forces in Germany (BRIXMIS for short), and their contribution over these four decades that undoubtedly contributed to the maintenance of a wider, but by no means ubiquitous, European peace that has endured from the end of the Second World War right up to today. It is also a personal perspective from my own brief sojourn, arguably bridging the most dramatic days, as the unit transitioned from the seemingly unrelenting pack ice of the Cold War to the bright spring of *Wende* – the drama of the Berlin Wall trembling and then collapsing into rubble – to the bright, bright sun of summer as the Soviet bear dramatically about-turned and began its trudge eastwards, rather than westwards, back towards the uncertainty of its own end of empire.

This book is not the first to be written about BRIXMIS – or 'the Mission' as it was more colloquially referred to – and there are other books published in English detailing life in the US Mission. What I aim to accomplish, against the wider backdrop, is the telling of some of my personal experiences as a professional intelligence officer serving in the Mission right at the very end of its life. Then, for the first time, give a flavour of the operations carried out by the Joint Intelligence Staff (Berlin), the son of BRIXMIS, once German reunification had precipitated the Mission's demise and most of its personnel had packed up and been posted home, until, in August 1994, the last Russian soldier, atop the last Russian tank, had left East Germany. I conclude with a brief taster, the coin flipped over, of what our Soviet opposite numbers, in their Missions based in West Germany, may have been up to – seemingly a very different but, to date, poorly documented story.

As a 'memoir' it is as honest and complete as I can make it. I have withheld some tales and experiences that should best remain untold, held back on some detail in others with an eye on security and in response to redactions requested during the process of security review by the Ministry of Defence but, frankly, these instances are few and debatable; I am talking about events that occurred more than thirty years ago, after all.

Operationally, BRIXMIS was not a hi-tech, high-value strategic intelligence unit, nor was it subordinated to any national agency. The equipment and tactics we used were basic and essentially extensions of the arsenal and fieldcraft of the infantry soldier. Put crudely and over-simply, but I think nonetheless succinctly, the espionage activities of the Mission revolved around a group of blokes, albeit highly trained and motivated, driving around a country the size of Cuba or Iceland or the state of Tennessee, searching for, then looking and watching and recording the military activity of our sworn Cold War enemies.

But, my colleagues and the generations who went before us did it very, very well.

If the sense emerges that there is a dearth of detail on the nitty-gritty of 'touring', I would agree. Again, frankly, there is only so much one can say about rattling around in the back of a jeep, with sometimes very little happening beyond the windshield, without it very quickly becoming repetitive and tedious – and it has been said elsewhere, achieving exactly the right blend and balance. What follows are the 'best bits' that remain fresh and vibrant in my memory – and thankfully I have always been blessed with a pretty good one. If I have recorded errors or made mistakes in recounting details, they are innocent and devoid of any intent. I am confident they don't alter the price of fish.

Finally, these pages also seek to give some impression and overview of how the Mission evolved from its first tentative baby steps in bombed-out, war-torn occupied eastern Germany to the unique, professional intelligence unit it became, and to elaborate on some of its key successes during those amazing forty-four years of operational activity. In so doing, I am aware that some of my own personal observations and conclusions may antagonise former Mission members. Comments about the ultimate contributions will not please everyone. But I sincerely believe I am only fleetingly reflecting what the recent release of new material and its publication suggests, and do not seek in any way to belittle, disparage or, worse, indulge in the ghastly modern, at best passive-aggressive and at worst wokist, trend to *rewrite* history – the amazing history of BRIXMIS, its twin siblings, and those worthy Cold War warriors who served within.

2

The Alpha: The Beginning

If you are younger than 40, I wager you have few personal memories of the decades we know as the Cold War. But was there ever a period in history when two enemies faced off against each other against such a backdrop shaped by mutual suspicion, distorted misunderstanding and tangible fear? Was there ever a period in history when mankind confronted the reality of mutually assured destruction and the end of the modern age – conceivably the end of history itself? There was: it was during these years.

If you are beyond the tender age of two score years, unless you are the very last of those Japanese soldiers still hiding out in a South East Asian jungle, you will have little difficulty recalling the trademarks, chapters and dialect of the Cold War. What about *sputnik*, Apollo, MiG, B-52, U-2, Polaris, SS-21, Cruise, Kalashnikov, KGB, GRU, CIA, *Stasi*, Checkpoint Charlie, Korean War, the space race, the arms race, the bomber gap, the missile gap, the re-scramble for post-independence Africa, the Caribbean missile stand-off, Vietnam, the invasions of the Soviet satellites, and the saga of Berlin – the airlift, the 'doughnut' speech, the Wall and its dispiriting tales of divided families and failed escapes? Some of the restrained language masking the potential death of hundreds of thousands will sound familiar: air bridge; Brezhnev Doctrine; DEFCON, 'defense readiness condition'; first strike capability; intercontinental ballistic missile; anti-ballistic missile; Star Wars; military-industrial complex; atom, hydrogen, and neutron bomb; nuclear deterrence; nuclear fallout; rollback; and the most chilling throwaway three words of the whole Cold War, in fact the scariest but most apt acronym in any language: MAD – mutually assured destruction.

But even if you recognise one key iconic symbol of the whole surreal story, the ideological division of Berlin, do you still remember how the carving up of post-war Germany looked? When I talk to my kids about the old West Germany, the former DDR and the sectors of Berlin, their return gaze is blank, and frankly their interest begins to quickly flag. When I explain not just the reality of a divided Berlin and the 43km of its Wall, its geometry of guard towers anchoring each and every 250m section of its length, the minefields and the vehicle traps, but the significance of this division, a capitalist western half, reflecting light, glitz, consumerism and personal freedom onto the drab, grey prison of socialism on the other side of *their* Wall, they struggle to grasp it. My words convey as much meaning and relevance as if I was miming in a blacked-out room.

Continue to elucidate on the wire that further divided 'mainland' Germany to the west of Berlin from the North Sea to the Czech border, and one might as well be speculating on the prospect of life on a Venusian moon. It is now ancient history set against the constraints and boundaries of our contemporary collective goldfish-like attention span and our unrestrained worship of *self*, and yet it was very real, the epitome of *Dr Strangelove* craziness: illogical, cynical and ultimately pointless driven waste. It did not, and still does not, make sense.

3

The Mission

This reality was, however, the genesis from which life was breathed into one of the most effective and longest-enduring intelligence units in modern military history: BRIXMIS – the British Commanders'-in-Chief Mission to the Soviet Forces in Germany, later rebranded, in response to a Soviet name-changing initiative, to British Commanders'-in-Chief Mission to the Western Group of Forces.

The Mission, as it was universally referred to by the relatively limited circle who were aware of the scope and scale of its roles, was an enduring product of both the alliance that defeated Nazi Germany and the ideological divergence that followed and defined the Cold War until its end.

By 1946, indeed earlier if you search carefully for the nascent cracks and fissures that permeated the anti-Nazi entente, it was clear to all that the wartime band of brothers was fracturing as distrust, fear and a wider and established, deeper-rooted geopolitical mould reasserted and reimposed itself to shape the new European, and global, reality.

Triggered by the precedent of the London Agreement of November 1944, an early prescient commentary on the perception of the growing problems attending this rebirth was the Robertson-Malinin Agreement, signed by the British and Russian military representatives who gave the document their names. It created a framework to guarantee the ease of continuing dialogue and liaison between the British military commander and the Red Army commander sitting in their respective post-conflict German headquarters in their respective military zones of influence within shattered, liberated Germany. It crafted a mechanism and means to 'jaw,

jaw', to misquote but anticipate the words of Churchill – a prototypical confidence-building measure. The wording of the three agreements – an American and French version followed hard on the heels of the British–Soviet original – ultimately ensured the longevity of the Allied Military Missions they created. The Missions were enabled to adopt new tactics and approaches, fundamentally to evolve into the intelligence units that they all became – if the agreements' wording neither permitted nor prohibited, that development was generally deemed acceptable.

In the key spirit of reciprocity that underpins diplomacy, the small bilateral military liaison staffs that the agreements gave birth to were guaranteed free and unobstructed movement within their respective zones to deliver messages on each and every day of every ensuing year, with no limit on the duration of this pragmatic, realistic and remarkably far-sighted requirement.

Interestingly, the original 1946 charter indirectly imposed a number of additional, broader tasks pertinent at the time for the first two or three years of BRIXMIS's activity. During this time the Mission was subordinated to the Control Commission Germany, and the team's members were actually paid directly by the Foreign and Commonwealth Office, sitting back in King Charles Street in Whitehall, until 1950 when 'ownership' was transferred to what would eventually become the Ministry of Defence. Consequently, personnel were charged with supporting the repatriation of prisoners of war, displaced persons and deserters, and as part of the Commission's de-Nazification programme, they played a part in tracing and extraditing war criminals. More widely, they were active registering war graves and helped adjudicate the settling of minor border disputes. And, perhaps surprisingly, Mission personnel were also mandated to feature in anti-black-market operations. These must have been challenging but professionally exciting and fulfilling times.

Accordingly, the British quasi-diplomatic staff was established, in preference to Karlshorst amidst the rubble of East Berlin, in Potsdam, in Soviet military territory in the south-western leafy suburbs of Berlin. Potsdam was selected as the location for the British Mission's official headquarters – although, in practical terms, Mission members lived in, and the de facto headquarters was situated in, the security of West Berlin – so it could be co-located with the Rear HQ of GSFG before it was moved to the former German *Wehrmacht* HQ and Army high command in Zossen Wünsdorf. A Russian team, known for ever afterwards as SOXMIS, the Soviet Military

Mission, was staffed and eventually housed in Bünde, within striking distance of the British military headquarters, in the heart of the British zone. The Soviets chose not to mirror the Allies' approach of operating from a safe base within their own territory. This is to say that the three Soviet Missions, operating in West Germany within each of the Allied zones of occupation, did not cross and re-cross the Inner German Border (IGB) from base locations within the DDR at the commencement and end of each tour. In the case of SOXMIS, they chose to be fully contained within their compound in Bunde, in contrast to the way the Allied Military Liaison Missions (AMLMs) crossed the Glienicke Bridge from West Berlin touring onwards from the Mission Houses in Potsdam. I suspect this approach was forced upon them, based on a pragmatic logistical assessment and the realities of time and distance.

Key to the agreement, and an enduring feature, was the maintenance of a system of 'passes', limiting the size of each team and effectively restricting the numbers who could transit through each respective military zone at any one time.* Critically, the passes guaranteed free, unobstructed passage to those Mission members who carried them. The British team received thirty-one passes for eleven officers and twenty additional supporting staff.

Identical reciprocal agreements, but with a smaller allocation of passes, between the Soviets and the US (the Huebner-Malinin Agreement establishing USMLM) and the French (Noiret-Malinin Agreement, FMLM) followed hard on the Soviet–British precedent the following year. In contrast, the US negotiated a mere fourteen passes, perhaps with, I suspect, a keener focus on the reciprocal number of Soviet officers this would allow to enjoy free movement in their zone, and the French eighteen. They too both formally established their Missions in Potsdam, with a back foot in their respective zones in West Berlin. This created a framework whereby British, American and French soldiers had near free access to what became, in 1948, the DDR, and Soviet military personnel roamed the respective zones of their former allies in the Bonn Republic, also known as the Federal Republic of Germany.

Complete freedom of movement was subsequently qualified in October 1951 by the imposition, in all four occupation zones, of PRAs, which guaranteed a degree of protection to areas – approximately 20 per cent

* I have included a copy of my pass in the photo section.

of each of the zones – that the hosts wished to remain off limits to casual terrestrial 'enemy' observation. These 'goose eggs' enclosed hundreds of square kilometres of choice military real estate – a combination of sensitive base locations and training areas. Those in the DDR were individually hand drawn onto small-scale maps and passed on to the Allied Missions by SERB (the Soviet External Relations Bureau), which provided, essentially, an administrative interface. By their very nature, these maps were a totally inaccurate tool and certainly not fit for purpose in adjudicating subsequent disputes as to whether a tour was in or out of a restricted area. It was, however, clearly understood that where PRAs landed on *Autobahns* (the world's first motorways), they could be transited by tours and, indeed, it was acceptable for a tour vehicle to halt on the roadside and for the crew to undertake observation tasks.

Fixing an upper limit on the percentage of territory that could be protected, this could, however, be augmented by limited-duration TRAs. TRAs were a convenient way of greatly increasing the size of areas out of bounds to tours by using them to link individual areas of PRAs when the integrity of a large deployment or training exercise had to be maintained.

A third form of restriction, which tours did not take seriously, has to be noted simply because there were so many of them: mission restriction signs. They had first come into existence also, in 1951 as local initiatives by Soviet commanders to limit Allied Mission access to their establishments and local areas, with absolutely no derived authority from the three original Mission agreements. The practice spread to cover the whole of GSFG, and indeed NVA installations. It is estimated that in 1964 there were well over 1,200 of them country-wide, almost as numerous as Hoxha's Dalek-like reinforced concrete pill boxes sown the length and breadth of Albania. By 1989 the number must have been even greater, but by then no one had the time or inclination to log them. Whilst they generated absolutely no protection to units sitting 'inside signs', they were fairly reliable indicators to a tour that had spotted them that they were approaching a barracks, some other form of installation or training area. New signs likely indicated a new target or, perhaps more likely, the fact they had been erected to replace old signs 'stolen' by tours as mementos and farewell reminders of life on tour. I regret not having 'half-inched' one, but it was another symptom of the 'oh, I'll get one next time' attitude like tomorrow, there is rarely the right next time – it never comes.

As tensions mounted and Cold War ice solidified into its state of seeming glacial permanence, the liaison role of the Military Missions, East and West, rapidly morphed into a new primary function that progressively grew and increased in relevance, becoming more and more focussed. Their liaison roles continued to play a vital role in de-escalating and managing tension at various key points during the freeze, but they evolved irreversibly into a wider function. The Missions became full-time dedicated intelligence-gathering units.

The Allied Missions' first intelligence task was to construct an Order of Battle for the Red Army units that were deployed in eastern Germany. This required teams to identify each and every unit at every formation level and trace their subordination in a meticulous and painstaking manner, filling the blank, virgin canvas. This intelligence was then briefed to the various intelligence staffs the other side of the Iron Curtain in western Germany, London, Paris, Washington DC and later in Brussels when NATO HQ relocated there from outside of Paris.

Tasking from Moscow to the three Soviet Missions, plus Berlin-centric city flag tours mounted through Checkpoint Charlie, I suspect, bore a few subtle variations, as I will later discuss, that cascaded down to our own Allied teams. This would have resulted, it was always consistently assessed, in at least a much more focussed agent-support function due to a more aggressive Soviet view of human intelligence (HUMINT) collection and SOXMIS staff's relatively easier and safer freedom of movement in western Germany vis-à-vis Soviet intelligence officers operating with diplomatic cover, or indeed as illegals.*

Although Russian xenophobia and paranoia, pre-dating 1917 and extending back into Czarist days, would have required their military teams to monitor, with greater or lesser intensity, Allied states of armed readiness, the western Missions performed this role with an enhanced existential and timorous focus. BRIXMIS, from the day of its first tour on 5 October 1946, and the other two Allied Missions, for the entirety of its operational life, played a vital role in NATO's early warning system. Its presence – 365 days

* There are few, if any, detailed and informed open-source treatises, that I am aware of, on the work of the Soviet Missions, but there are enough 'whispers' to begin to paint a much more nuanced output. Really weighty source material would necessarily have to come from within archives inside Russia. A fascinating project in waiting for an author or PhD student if these sources survived the collapse of the Soviet Union.

per year over the Wall, behind enemy lines, on the 'wrong' side of the Iron Curtain – was an integral and bespoke strand of the tripwire designed and resourced to warn the generals, and their political masters, of the Soviets' advance westwards towards the Channel ports, thereby, in design, giving them just enough time to execute their relentlessly rehearsed defensive plans to degrade and halt it.

If the wire was tripped, and successfully reacted to – a big 'if' of history that thankfully remained untested – the warning would likely, at least in part, have been triggered by the reconnaissance work of BRIXMIS and tours from the other Allied Missions. These tours were, certainly latterly, carefully coordinated by the Missions' operations staffs. The aim was to ensure there was a permanent presence monitoring the key units of the five Soviet ground armies and an air army – some twenty-four divisions – of GSFG, plus the additional eleven divisions of the NVA. The 300,000-plus Soviet conscripted troops of GSFG relentlessly trained within, and rotated through, this mammoth military machine. So, the doctrine dictated, if the Red Army was gearing up for that fateful and final advance, the Mission tours would not miss the intensification of activity associated with mobilisation. They might, indeed, be the sole Allied source reporting this fearful, fateful development. The stark reality was that such movement might not necessarily be detected by the strategic collection assets that traded in Signals intelligence (SIGINT) and Imagery intelligence (IMINT) – scrupulous radio discipline and bad weather could frustrate the Allied signals intelligence units listening in to GSFG radio and cypher traffic, and blind overhead collection by satellites or fixed-wing overflights.

The Missions were the unblinking eyes of the free West through every crisis that punctuated post-war history, from the sealing of Berlin – it was a BRIXMIS tour officer who reported the laying down of the first rough blocks of the Wall – to the ingress in 1968 of Red Army units over the Czech border from the deep south of East Germany, to the invasion of Afghanistan, right up to that potentially destabilising November evening, and the weeks and months following, when the Wall so dramatically came tumbling down.

A fundamental complementary function of this role, observing Soviet military preparedness, was the unrivalled opportunity to gather a whole cross-section of military and infrastructure-related intelligence. This involved monitoring the unending cycle of training exercises from the initial low-level training, marking the arrival of the latest influx of raw conscripts every six months, to the culmination of their two-year tours of duty in the divisional and Army-level combined large set-piece exercises. It encompassed commenting on levels of competence of both ground and air forces, assessing tactical doctrine, identifying and evaluating new equipment in the Soviet arsenal and recovering a mountain of technical intelligence that few other collection platforms and agencies had the opportunity or potential to gather. In infrastructure terms, teams were able to bring west *de visu* intelligence reporting on everything from the DDR's rail network and uranium ore mining to the span of key bridges and the state of repair and implied capability of other key pieces of infrastructure to the identification of auxiliary runways routinely engineered into the *Autobahn* network. So simple, so direct, so reliable, so unbelievably cost effective, but potentially strategically priceless to the Western alliance.

Significantly, by way of a footnote, the post-war Mission model lived on beyond its own demise in 1989. It served to influence the various arms control negotiations, and shape the resultant teams that were configured to monitor the specific and knotty detail of their agreements, playing their full part in keeping the world a safer place.

4

Training – Ours Not Theirs

My arrival in the Mission was yet another fortuitous twist of fate, benevolent karma, that had characterised the eight years of my career since leaving university. In fact, I should include those three years of study in this calculation as, from my first year at university, a martial god had smiled on me, pushing, prodding, leading, then preparing me to secure my Cadetship. This had meant all my university fees had been paid by the Army, plus a salary as a second lieutenant had flowed monthly into my bank account – I have never been wealthier. And certainly my father was delighted: no requirement to bankroll this son. As I looked back at the almost casual, certainly unorthodox, way I had slipped into the Special Duties world, on the back of that first posting to 123 Intelligence Section in Lisburn in Northern Ireland, then my three tours of duty in the Force Research Unit running covert agents against the terrorist groupings in Northern Ireland, my charmed destiny-driven luck had continued with equal, almost unbelievable, good fortune.*

In my early meetings with Ray, before I formally handed over South Detachment, Force Research Unit (FRU) to him, he had talked about his days 'in the Mission' in Berlin touring around eastern Germany watching the Russians. It was hard to believe this was an operation that we, the British Army, were engaged in and had been for so long. As Ray talked more, I was struck by what a well-kept and enticing secret the real role of this extraordinary unit appeared to be. He spoke about shooting incidents,

* These experiences are related in *The Deadly Game*.

rammings, and other rough-house stuff, racing at high speed to avoid pursuit by the *Stasi* or Russian military vehicles and regular contact, not just with ordinary Russian officers and soldiers but with uniformed members of the KGB and GRU, Soviet military intelligence. Tales of the day-in and day-out operation to collect intelligence against these sorts of challenges represented exactly my idea of a good posting and the next job I needed after the high-octane life of a FRU Detachment commander. I had heard the name BRIXMIS during my attachment to 3 Intelligence and Security Company, stationed in Berlin, during my final long summer break once I had sat my university finals, but I had had no clear understanding of the unit's role. All I had really known was that it was a very sensitive unit and was considered to be another tool in the 'Special Duties box' that contained other units like the FRU and 14 Coy across the water. Ray made it sound very, very interesting indeed to the low-boredom threshold and restlessness of my character.

I had been looking at the prospect of either remaining in Northern Ireland in an intelligence liaison role with the Royal Ulster Constabulary (RUC) Special Branch or returning to Specialist Intelligence Wing to run the training that supplied the FRU with its operators, back in Ashford. In the event, I had fallen out of sync with these openings when I decided to take the opportunity to return to the FRU for one final year as the unit's intelligence officer, a new position basically controlling the flow of intelligence within the headquarters and its dissemination out of the unit to our strategic-level partners – essentially, but not exclusively, to the upper levels of the Special Branch. Against this reality, when the Intelligence Corps' officer manning desk mentioned the prospect of a posting to Berlin and Ray's old unit, I bit off their hand so quickly and cleanly there was not a single drop of blood to spoil the worn carpet.

I have reflected so many times on the perversity of fate – my choice, following Ray's advice, leading me to yet more adventure against unorthodox and irregular backdrops; Ray's decision to serve with the FRU ultimately leading him to his tragic and untimely death in the Chinook crash on the rugged Galloway coast of Scotland six years later in 1994, along with twenty-eight other military, police and civilian intelligence officers.

As I pulled up and parked below London Block, the imposing, utilitarian Nazi-era sports administration building within the 'Olympic Stadium' complex that was the Mission's West Berlin home – so named, without irony or, indeed, imagination, because the 1936 stadium was just a javelin's throw across the playing fields – I had absolutely no comprehension of what the next three years would bring. I was certainly thrilled and excited to be back on the island that was Berlin, looking up to the stone eagles that had kept a watchful eye on BRIXMIS all these years. How tumultuous these coming years would be with the re-alignment in European, and ultimately world, history, and how the role of the Mission would change in so few months from my applying the handbrake on my BMW, were well beyond the limits of my imagination and, I am positive, the occult powers of others.

I had joined the Special Duties course run by Attaché Branch based in Templer Barracks in Ashford, Kent, the home of the British Army's specialist cadre of professional intelligence officers, the Intelligence Corps, earlier that summer. The year was 1989, seemingly unremarkable but how portentous a date.

Attaché Branch schooled our military attachés on fulfilling whatever intelligence tasking they would be mandated to fulfil in-post. Clearly this was little in the case of those going to Washington DC, Ottawa or Canberra, but a little more onerous in the case of Moscow or Beijing, or the capitals of strategically placed developing states.

I had already had a couple of weeks' leave to decompress after the numbing stress of my time as Intelligence Officer FRU and was readily anticipating the challenge and variety of something new after nearly six years' almost unbroken service in Northern Ireland.

I thought the course title was rather grandiose and stretching a point as, in my book, Special Duties meant service in the Province, after tough selection and training with either 14 Coy or the FRU. There certainly was no element of competitive selection that was apparent to me for Mission service and, as far as I could make out, no one ever failed the course. I preferred to refer to the training as the BRIXMIS course. Anyway, moot points, I suspect. In reality, the fact was that BRIXMIS was by then already forty-three years old and had trademarked the term long before. Leaving unit snobbery and arrogance behind, I concentrated on the job in hand.

I am sure I am right in saying there were still two courses run per year. They now lasted four weeks, but earlier courses dating from their inception in 1972 had been shorter. Even though members from the other two Missions blistered onto our training en route to Berlin – there was agreement to put four USMLM students every year on the course – the throughput in BRIXMIS did not require a larger number of new personnel to be trained, and Attaché Branch had other core 'customers'. By the time the last training package had run its course the following year, a total of forty-nine had been run. That is an impressive record.

We were, therefore, the next and newest generation of 'tourers' in every sense of the word. We were from an eclectic mix of backgrounds. There was Martin, a fellow Int Corps officer and Russian interpreter – we became even closer friends in post-Army days as we both worked for Philip Morris International and later on a project based in DR Congo; Steve Gibson, a good friend and a sound operator with much too keen a brain for intelligence work, or indeed for the Army, who went on to write the definitive book on Mission touring and was a key figure in writing up the official history of BRIXMIS; Gary, an ex-Royal Marines SNCO who had passed selection and was badged SAS; Alec, an ex-Guards SNCO who had a long pedigree with 14 Coy and was what was referred to as 'permanent cadre'; Sinclair, an Int Corps SNCO who came to work directly for me; a handful of others including the newest crop of Royal Corps of Transport drivers; and Jim and Bob, two US officers who were bound for USMLM. Jim was a US Marine Corps officer. He had joined as a weapons officer/navigator flying on Phantoms but had had to bale on that because he suffered from incurable air sickness. And Bob was a US Army officer. I became good friends with both, as much as you can with Americans – and I certainly do not mean that in any pejorative sense, just a feeling I have; I think you only ever get truly close if you marry one and this was never an option. Once settled in Berlin, we socialised together, and I enjoyed the fresh ideas and perspectives each brought to military and civil life there. As a wider point, I always enjoyed every contact with our American opposite numbers, work or socially related, and took every opportunity to understand their operational and tradecraft approaches and compare and contrast them to our own.

Everyone on the course was a good guy. In fact, just about everyone in the Mission, including USMLM, was a decent human being and good to be with at work or play. Given language limitations – theirs, of course, not mine – I did not get to know any of my French opposite numbers that well, but they always struck me as very solid and professional, getting each job done and moving on to the next without fuss or drama.*

The course was intensive, hard work, but lots of fun. Like the courses run across the parade square at the 'Manor' in the Specialist Intelligence Wing (SIW), there was no bullshit, no 'us and them' posturing between directing staff and students, just a positive sense of professionalism and purpose. The attitude was 'We know where you are going, what you will be doing, and we will prepare you as best we can so you can accomplish those tasks.' In my book, there is no other way to conduct or structure training; training is strictly functional with one aim and one aim only. I knew most of the teaching staff from periodic visits to SIW back from the FRU – Templer Barracks was always a small, intimate place. They shared backgrounds either in BRIXMIS or as former defence attaché staff. One of the RAF officers had served on the air attaché staff in Moscow and had flown one of the fleet of Victor tankers that had refuelled the Vulcan bomber that had successfully bombed Port Stanley airfield during the Falklands War, dealing such a powerful psychological blow to the Argentine invaders – a truly amazing operation.†

The first part of the course was entirely classroom based. There was an incredible amount of 'death by slide' simply because we had to learn to recognise the whole Soviet arsenal, main variants included (and the Russians loved variants) literally every piece of equipment that was deployed by both the Soviets and the East German NVA in the DDR, and every system that could be moved into theatre, even if only on a temporary basis. This included everything that flew, whether it was a fighter, a bomber, a transporter, a helicopter or anything else. It included every ground-to-air missile, its launcher and associated missile trans-loader, radar, and every rocket launcher and every rocket, including the range of nuclear-capable

* I remain uncertain as to their relationship, if any, with DGSE, the French secret intelligence service. I would work more closely with the French Army in Bosnia when, post-Dayton, my HUMINT team packed away their UN berets and re-roled under NATO IFOR command. I had a French captain joining the team from the RDP, a mobile special force that specialised in battlefield surveillance. He was a highly experienced soldier and a really solid and dependable team player. We worked alongside his parent unit, providing them with operational intelligence that on one occasion, early in 1996, led to the arrest of the first Islamic foreign fighters supporting the Bosniak army, the ARBiH. We had, unknowingly at the time, uncovered the tip of a *Titanic-sized* iceberg.

† Rowland White's *Vulcan 607* deconstructs the incredible complexity of the mission beautifully. The operation was a stunningly brave and ambitiously courageous achievement.

systems. There was a host of artillery pieces to get under one's belt, not just towed guns but self-propelled tracked ones and anti-aircraft weapons. Then, of course, there were all the various tanks and tracked and wheeled armoured personnel carriers (APCs) that a tour might see deployed out and about in GSFG's tactical area of responsibility, communications equipment and communications intercept equipment. And then, just when you thought you had won, a host of trucks – big trucks, small trucks, long trucks, short trucks, trucks with trailers and without – cranes, vans, jeeps et al. Literally hundreds of bits of 'kit', a great word and so evocative in the unit – a tour would spring into action at any sighting of equipment spurred by the battle cry of 'Kit!' We even looked at some ships because Mission tasking also included some naval targets up on the north coast centred around infamous places like Peenemünde where the V1s of the Second World War had been tested. To spice the process we were also subjected to the subtleties of 'tarpology', studying and learning the outlines and profiles of equipment being moved under the cover of canvas tarpaulins to enhance their security and environmental protection. This was obviously a less exact science but a necessary skill if our collection efforts were to be maximised. To add realism, we were tested on long-range views of equipment, close-up detail and views that were partial and involved significant parts of the system being obscured or concealed – just like they would be out on a cold, wet, foggy, grey training area with light levels dropping and dusk descending.*

Beyond equipment recognition, we were taught the significance of a sighting of a particular piece of equipment, not just in isolation but what it was that the sighting signified in terms of its subordination to a formation. Why was it there, what was it doing, and what might be happening around and beyond it? We needed to be able to furnish instant answers. We had to understand at what formational level these weapons systems were operated, and the import of certain sightings and combination of sightings. Ultimately, the question we constantly posed out on the ground sitting comfortably, or uncomfortably, in our G-Wagens was, was there scope and

---

* I remember idly calculating just how many images we had to force our memories to juggle with. The total was not far short of 1,000. To make the process of remembering more challenging, we had to be able to identify the equipment instantly, times table-like, without a moment's pause, or it would have been impossible, when touring, to tackle moving convoys or kit trains effectively.

requirement to push up against and beyond the routine collection envelope? This always brought us back to the quintessential challenge of assessing risk versus gain. I have never liked the term 'expert', but everyone had to become, of necessity, very, very quietly and demonstrably proficient.

Normally, the prospect and reality of being sealed inside a military classroom hour after hour and day after day, tomb-like, would be met with at least raised eyebrows if not declarations of intent to go AWOL, absent without leave. Frankly, however, I enjoyed the novelty. It was just pleasant to be sitting down comfortably in a warm, dry classroom. I found it therapeutic, after the nuances, subtleties, stress and complexities of agent-handling work, to be either unequivocally right or unequivocally wrong with absolutely no shadings of grey or room for comment, debate or interpretation. Either it was a 2S-3 or it wasn't. I enjoyed the challenge of assimilating this massive amount of data and then demonstrating that knowledge, pub quiz-like.† I also appreciated the novelty here getting up every morning. It was a bit of a pain shaving every day, but in contrast to having to decide whether today it was this pair of jeans or that pair of jeans, this T-shirt, that T-shirt or that shirt or jumper, it was so straightforward to put on those same green lightweight trousers, that same green combat shirt, that identical green woolly-pully with those green socks and those black combat boots before heading down to breakfast.

With this arduous initial phase over, the practical part of the course commenced. It dealt with photography, the skill that underpinned everything the Missions did, the tactics of touring and the operational skills required to tour effectively, securely and efficiently to collect the intelligence we would be tasked to gather once on mission in the east.‡

---

† We continued the training and testing once in the Mission with weekly sessions kicking off Wednesday mornings, rather like morning assembly at school, or perhaps more aptly the weekly maths test. We all took turns to put together a box of equipment slides, which the assembled Mission would 'spot' and identify, image by image. Shots deliberately and pointedly included incomplete profiles, tricky and ambiguous angles, lookalikes, any subject or perspective that could test a tourer in the field. Failure to maintain a sufficiently high score did not mean detention the following evening; it was rather more serious than that. Failure to make the grade meant operational grounding – so, mind-focussing stuff and a big incentive to keep up to date and familiar with the hefty recognition pamphlets that informed the ritual of our personal prayers.

‡ Interestingly, from 1964 the Americans had received their photographic training at Leica's Photographic School in Wetzlar in West Germany. I am not aware of the date this approach ceased but it must have been a novel and rich experience and no doubt imparted the highest level of instruction and knowledge.

I think I took things a little for granted because my short career to date had been entirely operationally focussed. Where some of my fellow students might have faced this phase of the course with some trepidation, I had already spent a year in Northern Ireland with the infantry operating in County Armagh, as well as my time with the FRU, and I'd completed a long covert surveillance course. Hopefully without sounding arrogant, I was fortunate to be confident out on the ground with enough experience under my belt to handle most intelligence collection challenges.

So we made Ashford our Potsdam and toured out against 'enemy' targets to our west and north and east. We did not need the Red Army or the NVA. We hit Army static installations and exercise played on and around Salisbury Plain, including the big garrison towns of Bulford, Warminster and Tidworth. We 'attacked' the aircraft test site at Boscombe Down and, slightly more chillingly, the Defence Science and Technology facility at Porton Down. And we toured against NATO air targets on the flatlands of East Anglia.

The key lesson the newcomers to clandestine operations learned very early was the requirement and supreme importance to 'know the ground' – this maxim underpins operations whether they be agent handling jobs or surveillance tasks or the reconnaissance tours that we were to mount. A sound knowledge of the geography gives an enhanced edge to operational efficacy. Without the need to be constantly head-down in a map book meant one could capture maximum detail and maximise intelligence collection, but beyond that, it increased situation awareness, which directly translated into enhanced security. It was unlikely we would draw enemy fire here, but that was certainly not a given where we were all going in only a matter of weeks. That said, however, the ability to map read well was essential, but when close in to a target, there was no substitute for feeling at home – I have had the conversation multiple times when tourers have stated, without embellishment or exaggeration, that they got to know some areas of the DDR better than those around their own homes. Nothing beats the experience, in this context, of having visited a location before, but without this luxury, proper preparation was paramount. When we arrived in Berlin, preparation for each tour would entail detailed study of the target packs that existed and had been compiled and amended over the long years of previous intelligence attacks. This confidence and ease distinguished the good tourer and the good tour from the average.

The reality of Mission life was that there were of course no allusions on either side as to what each other's 'liaison staffs' were really doing. We knew that SOXMIS and the other two Soviet Missions in western Germany were engaged in intelligence collection, and the Soviet forces of GSFG and native DDR formations of the NVA of course knew the French, US and British Missions were mounting collection efforts against their armed forces and military installations. The name of the game was to not be caught flagrantly, especially on film, in the act of collecting this intelligence. The aim was to refrain from taking undue interest in a military target when this could be observed, especially by a Soviet or East German NVA, military 'bystander', a *Stasi* surveillance operator or even a member of the public – we presupposed, worst case, that all civilians, as good SED party supporters, might report our activities to the authorities.

This necessitated tours to employ stealth as their default guiding principle when tackling a target. Apart from the requirement to remain 'downwind', these were essentially the tactics of the hunter. We were stalkers, in the traditional, purist sense. We practised use of covered approaches, minimising noise and observing from cover, invariably from a position standing off from the target. We made full use of camouflage, the most practical manifestation being that the Mercedes G-Wagen could be easily mistaken for the ubiquitous UAZ-469 Russian jeep. But when the situation required it, a final high-speed, aggressive closing with the enemy, then rapid withdrawal was not an uncommon tactic.*

Course exercises tried to mirror as closely as possible the profile of touring for real. In the case of the longer ones, at the end of each day the crews would cook and then bed down in a wood or copse for a few hours of sleep before the next day's exercise serials. It was not important to extend activity beyond a single overnighter, so we were never away from Ashford for too long.

In the event the Soviets considered a tour vehicle was paying undue attention to their presence, they could 'detain' it. This meant essentially 'arresting' the vehicle and its occupants. Detentions could be rough and tense and might

---

* If the tactics of touring interest you, read Steve Gibson's really excellent *The Last Cold War Mission* – it captures the blow-by-blow, move-by-move drama of touring better than anything else written about the Allied Missions. I can fully recommend it. As the Americans say, 'He nailed it'. We served together and I can attest to his professionalism and energy. Steve scored a key goal when he recorded the first sighting of an SA-10 in the DDR.

entail a tarpaulin being thrown over the vehicle to deprive the occupants of any further view of their immediate surroundings, and to intimidate them. Let's not forget that all our various contacts with the soldiery and citizenry within the DDR were underpinned by the very real reality that we were sworn enemies and the Cold War was ultimately a potential struggle to the death of two civilisations with fundamentally opposed philosophies and approaches to all aspects of life. However, detentions would only be of relatively limited duration, and tensions at this micro level would always eventually subside; the brief to teams was to refrain from taking any action that would escalate matters. Observation equipment–cameras, binoculars, night-vision devices–mini tape recorders, map books and all the other tools of the trade would be packed away, and tourers would assume the attitude of disinterested 'tourists' or, rather more pertinently, fully accredited and official 'liaison officers'. If vehicle doors remained locked, there was nothing the Soviets could do to compromise the tour members, but there had been occasions right up to the recent past when tourers had been pulled out of vehicles and beaten, and equipment had been grabbed and taken. Ultimately, tempers and reactions were governed and directed by the cover the Missions enjoyed as diplomatic entities, but in the same way that diplomats can be *persona non grata* (PNG), particularly effective tour officers had been targeted for repeat detentions so their expulsion could be justified. Essentially, governing relations was the appreciation that the treatment received by one side's Mission's tour would be meted out in reciprocity to the other.

In most cases detentions created opportunities for contact and low-level liaison between the two sides. Eventually, the *Kommandant* from the nearest *Kommandatura* would arrive to take charge of the incident. He would attempt to get the tour to sign an '*Akt*' admitting they had been engaged in improper activity and/or were in an area where they were denied access, which of course tour officers refused to sign. All Mission crews carried detailed mapping accurately displaying PRA borders, which were religiously and promptly updated when revisions were made. Invariably, the Soviets would argue the case for Mission tours to be in a PRA on ridiculously small-scale sheets upon which the width of a pencil line represented several kilometres. These were invariably uneven debates. By implication, however, the very existence of the *Akt* suggested that on some occasions they were, or had been, treated with respect and a signature, but certainly by 1989, no more. Eventually, the tour would be released, generally after an imposed rest period of several hours, and let go, to continue its collection mission – all a merry little dance.

Long story short, we were subjected to a detention on the course at some stage during one of the exercises, complete with character players dressed in Russian uniforms and with exchanges in Russian. The charade was a useful little tutorial but provoked none of the real-life angst and drama that could accompany the real thing. I was detained later in Halle and I have to say the lessons identified by our mock arrest did prove useful, but ultimately the knowledge that, barring a freak accident, you were not going to be hurt or unduly stressed made detentions little more than interesting and novel distractions. A few years later during the Kosovo War, I was in the far north of Albania, heading up to the border with Kosovo and was hijacked by heavily armed bandits. This was a rather different deal, a little more stressful and a little less controlled and predictable in its outcome. But the same basic approach ensured this experience had an equally happy ending. An important lesson of life is always to attempt to engage an aggressor in conversation, the calmer and more rational the better. Someone talking to you is unlikely to be someone engaging in simultaneous acts of violence against your person.

The ground tours we practised gave me the opportunity to see a bit more of my Army, but the most interesting to me were our runs to target the USAF air bases up at Lakenheath and Mildenhall. Air tours complemented the Missions' ground-related activity. They too were terrestrially mounted, but not exclusively as I will explain later. The 'air' aspect of their nature meant targeting aircraft, radar and air defence equipment. Against Soviet Air Force targets, this related to observing both the tactical deployment of systems and the deployment of new equipment, typically underslung bombs and missiles, newly deployed radars and associated hardware and surface-to-air missiles (SAMs). Resultant photography of course identified and confirmed new kit but it had to be of a quality and resolution that allowed analysts to extract technical intelligence. This entailed tours targeting Soviet aircraft by attempting to get in position at some point along an airfield's centre line to observe and photograph the underside of the planes on final approach, coming in to land, when flight speeds were at a minimum and the configuration for landing provided ideal observation opportunities.

As a lifelong air enthusiast, this was an exciting and visceral place to be as these USAF jets thundered in just above our heads, wheels down in all their glory. There is nothing like the power and scream of aero engines, the wake turbulence and just the sheer spectacle and awe of man in flight. With the right shutter speed and correct degree of over-exposure, every rivet and bolt popped out in stunning detail onto the prints once they were developed – those near forgotten days of pre-digital wet film.

The course ended in the same positive and consistently professional spirit in which it had been conducted throughout. I know for sure that our American counterparts particularly enjoyed this taste of 'Englishness': their stay in a quintessentially British officers' mess and the view they enjoyed of a less well-known part of the British Army at closer hold. The close proximity of Ashford to London was an added bonus that didn't hurt. With everything packed up, including an extra suitcase full of our newly acquired recognition pamphlets, only one adventure remained before our arrival at the Olympic Stadium and the prospect of a full and exciting tour with the Mission.

5

The Road to Berlin

The eight-and-a-half-hour journey from Calais to the small German town of Helmstedt, sitting just shy of the Inner German Border, was long but uneventful. I'm not sure it has ever been described as an interesting or exciting route since the demise of the stage coach. But it was interesting to compare a composite of the various driving characteristics as one traversed the three neighbouring countries. French drivers are fast and, one senses, never too far removed from reckless catastrophe. Belgian drivers are generally poor, unpredictable and seemingly distracted, perhaps by the ever existential concerns about the unity and longevity of their country, Jacky Ickx exempted. German drivers, Teutonic to the core, are fast, aggressive but skilled and predictable, and their system of *Autobahns*, this side of the ideological divide, was, and remains, superb despite the contemporary chaos of ubiquitous sections of roadworks seemingly every 50km.

In dramatic contrast, the next leg of my route was anything but tedious, routine or forgettable. I saw too few drivers to comment on their levels of skill, and anyway I was persistently distracted by the unrelenting presence of the Trabant car. I very quickly accepted the normality of their existence but there was no greater reinforcement of the divide between East and West than the contrast between the small span of motorway between Hanover to the IGB with its hungry BMWs, Audis, Mercs, Opels and every other symbol of the Western way of life, and the run-up from Helmstedt to the capital of the DDR, with 'Noddy' car after Noddy car, with their other-world colour jobs, interspersed with an occasional Wartburg or Lada. The odd blue and beige shades of the Trabi appeared nowhere else in the palette of either nature or the man-made world of consumerism.*

Helmstedt represented something different and an utterly surreal break in the route to Berlin's Dreilinden. This, point to point, was the 170km stretch across the flat north German plains between the officially christened Checkpoint Alpha and Checkpoint Bravo. Helmstedt not only marked the frontier between free West and communist East, but was the 'gate' beyond which one of the most novel, memorable and totally unique road journeys in the world began. This small, otherwise insignificant, Lower Saxon, formerly Hanseatic, town was the most westerly point of the 'corridor' to the island of Berlin, marooned in the heart of the German Democratic Republic. This was the single transit road route into West Berlin, and beyond to Poland, from the British zone of the Federal Republic. I think I am right in saying that both US and French servicemen may also have used the corridor. In real terms, it was a bridge, a causeway rising above the quicksand of the East – remain on it or fall off the world into the abyss.

To access this northern route, all travellers had to book in through the communist authorities. In the case of non-military travellers, including West German nationals, they were processed by an East German control system. British military servicemen, however, were shepherded differently, in line with the post-war system that underpinned everything to do with the east. We had to initially pass through a Royal Military Police (RMP) checkpoint, where we received a rapid but serious and slightly intimidating briefing on the next leg of our journey. Here I was informed I had to maintain a constant speed of 60mph, no more and no less. My journey time would be monitored and I would be punished for an arrival at the

* The Trabi, or 'companion', was a marvellous testament to the socialist ideal. Just over three and a half million were manufactured over the thirty-four years it was produced. During this time the basic design evolved into a mere four variants. How wasteful is the West, in comparison, with its ever-changing models and marks. Trabi bodywork was made of Duroplast, a plastic made predominantly from recycled cotton from the Soviet Union – an environmental approach way in advance of our contemporary Western preoccupation with recycling. Its two-cylinder, two-stroke engine produced 26hp and could get the vehicle up to 60mph/100kph in a sprightly 21 seconds – stepped-down performance clearly with a view to enhancing road safety and economy, unlike our Western thirst for power and speed and ever greater performance. The only downside I can see with the Trabi was its very poor emissions record. But it has to be said that its impact on global warming was minimal, as there were relatively few 'easties' who could afford one. And this situation has markedly and dramatically improved – the Trabi is now an ecological star: no one makes new ones, no one buys old ones and very few drivers drive any version. A treasure in anyone's book.

other end of this 'tunnel' that was too fast or too slow. There was to be no stopping for any call of nature involving either end of the alimentary canal or, for that matter, including the urinary tract. Deviating from this section of *Autobahn* was *verboten*, and in the unavoidable event of an accident or breakdown, I was to remain, without fail, with my vehicle until RMP recovery arrived. Frankly, as a newbie I did find this degree of gravity a bit unnerving and unsettling. But by my second journey I was much less intimidated. By my third, I had tired of the formality and constraint and resented the imposition of this unavoidable brake that dramatically slowed journey speeds and increased journey times – how quickly we fickle humans take even the most noteworthy events and objects for granted.

Once clear of the RMP, universally referred to as 'monkeys', the true odyssey began. As we did not recognise the authority or legitimacy of the East German state, or any of its institutions or its public servants, we eased carefully through the barriers, wire and Soviet guard towers, arriving at the Russian military checkpoint. This was a graphic reminder that we were now very much in their backyard, about to go through their garden gate, and that they were the big boys on this block. Parking up and surveying this alien scene, I got out of my car, locked the door and advanced to present my paperwork to support my right to cross the Soviet zone of control to the free city of West Berlin. Of course it was clear that nothing untoward would, or even could, realistically happen but I always experience an unavoidable quickening of the pulse crossing any frontier or passing through any immigration or customs control. I entered the small 'guardroom'. This was now the closest I had ever been to the enemy, despite seeing Soviet personnel crossing through Checkpoint Charlie, from East Berlin into West Berlin, on my post-university attachment to 3 Intelligence and Security Company eight years previously. I was now amongst them: real Red Army soldiers. I remember that the guardroom had a grotty toilet for our convenience – I passed on that one, even remembering the strict 'no stopping' regime. My bladder would just have to expand. I handed my paperwork across the counter and waited whilst the Soviet officer disinterestedly checked it, but no doubt making a log entry for their security/intelligence staff somewhere in the system to collate. I survived this routine scrutiny and, papers back in hand, breathing a little more easily regardless, retraced my steps back to my car and set out on my heavily regulated journey.

It turned out to be both quiet and uneventful. Traffic was light but there was my first confirmed sighting of a Trabi – a personal scoop, Mission parlance for a significant find or sighting dating from the Annual Report of 1984 but forever after used to denote the capture of a new Soviet armament. I witnessed a few Soviet trucks, but no exciting equipment, routing through the infamous 3 Shock Army tactical area of responsibility, or TAOR in mil-speak, that required me to access any of the recognition guides indelibly etched and shelved in my memory. I managed to successfully navigate the only point of potential consternation where the *Autobahn* split, taking the unwary south towards the DDR heartland and trouble. And then I had passed seamlessly through Checkpoint Bravo back into 1989, emerging from the frozen past that the DDR represented, returning to the glitz and frenetic pace of tomorrow's capitalist west, looking forward to the next time I would be 'through the looking glass' on an official BRIXMIS reconnaissance tour.*

* Looking to the not-too-distant future, in contrast, one of my good friends from the Mission happened to have to make the return journey westwards just after the Wall had come down. It took him an incredible eighteen hours to cover the distance to Helmstedt, so many easterners were grabbing their god-sent opportunity to cross into the west and taste their first bite of freedom in the other doppelgänger Germany.

6

BRIXMIS

I quickly settled into the Mission and into the rhythm of life in the unit. I wouldn't say the welcome was effusive but it certainly was not negative or intimidating. It was never easy being posted to a new place and a new organisation. The very process of rolling up life in one place, wrapping and packing it up, suspending or putting on hold friendships and, for those with family, uprooting kids from school and their network of friends and activities, ripping spouses from jobs and potentially diverting career paths, anticipating a new home, maybe a new culture, new infrastructure, perhaps having to adapt to a new enhanced security regime and all the other little things that weave themselves into a life was never stress free but, in counterbalance, rarely a totally negative experience. If one could not manage it readily, then Army life certainly became impossible. Moving to a place like Berlin, however, definitely qualified as one of the best postings one could land, regardless of unit.

Joining a new unit meant you were the outsider coming into an established environment in which friendships and professional relationships were already established, and unit members were familiar with processes, procedures and ways of doing business and getting the job done. These lessons one had to learn quickly, and one had to concentrate and focus to avoid making early mistakes in the process. Human nature is such that the outsider, and by extension the newcomer, is at least passively resented and invariably received with some reticence. This tended to be a short-lived phenomenon when the new man was professional and able to project positive human qualities. This was always one good thing about the Int Corps in that, because it was relatively small, paths would cross and re-cross; colleagues were at least friends of friends or were known quantities by reputation – this was especially so on the Special Duties side. I have to say I was never really overly concerned about any of the peripheral factors of moving on posting nor the professional challenges. This was partly due to my experience from childhood growing up in a military family – disregarding the old chestnut that my father's Royal Air Force service did not qualify as belonging to a military organisation – and unquestioned acceptance of the reality of regular cyclical moves, but also my philosophical approach to life. I was a firm adherent of the belief that fear generated by anticipation of, and speculation about, specific hurdles or challenges lurking waiting to be confronted, or indeed the future in general, was never matched by tackling the reality of a situation. The move to the Mission, even for those anxious and unsure of their new reality, was made easier by the fact the group of us who had been on the course at Ashford arrived en masse – we had already generated a group bond and established close working relationships, if not nascent friendships.

The unit had an air of permanence about it, mirroring the divide of the city of Berlin and the two Germanys themselves that had given birth to it – a well-established, well-tuned machine with over thirty years of solid track laid behind it and a clear path leading, seemingly, into an unbroken future. It had almost the feel of continuity, tradition and mutual shared identity that an infantry or cavalry regiment shares. As individuals we were merely interlopers in something that was far bigger, and well tried, tested and focussed by the heat and passage of past deeds and the triumphs of past members.

Tangentially, one thing I had noticed consistently during my career was the attitude towards our professional forebears and their exploits. I am not sure whether this was a phenomenon limited to the world of intelligence gathering or whether it permeated other military walks of life. I sensed that certainly in the infantry the issue of battle honours and shared former glories was a serious cultural issue, and those who had fought in past wars were honoured and respected positively. However pervasive, *our* attitude was not pretty. In fact, it was ugly and pejorative. I know I had been equally guilty and negative, but there was a strong and powerful tendency to regard the efforts of those who had gone before us, not just combatants in other wars but merely, taking the case of the Northern Ireland emergency, the generation serving before us, in such a way as to minimise and diminish not just their achievements but the very way they operated and got things done. In our minds, we viewed them as amateurs and their tradecraft amateurish, bereft of the subtlety and sophistication of our own efforts. We were the professionals: sharp, better trained, better equipped and with a better understanding of the mission. Of course, a coin does have two sides. To that generation that we held in professional contempt, from their perspective looking forward to our exploits, *we* were the amateurs. We were the new kids on the block, the young pups, lacking their professional gravitas and hard tried and tested experience. This is certainly the way I regarded those members of the FRU who followed the teams who served with me. And I have to admit that I was professionally dismissive of their technical and tradecraft abilities and performance – but certainly not their personal courage – and the quality of the intelligence this new generation of agent handlers brought to the conflicts in Iraq and Afghanistan; I feel I can say this as I oversaw the effort in the early days of the stabilisation operation in Iraq.

Perverse approaches to life, but undoubtedly every military generation experienced an enhancement of technical resources and capabilities which, certainly in my time, made us more effective and enhanced our security, but were we so much 'better'? With the benefit of time and distance to guide reflection, I think not: these attitudes were born of ignorance, arrogance and hubris, but were no less real for it. Looking back, how stupid and divisive a perspective – but how human.

I did not, however, regard the exploits of those who had occupied the Mission corridor before me in any derogatory or reductive manner. I had studied the Cold War at university and had continued to increase my knowledge of east–west relations ever since, so I knew the pressures the Missions had faced and that the tensions that had guided and shaped their operational environment had required both skill and resilience in order to meet, and master, the challenges. I saw myself as here in Berlin to continue this proud and effective tradition.

I was posted into the Mission as what was referred to as a 'part-time tourer'. My posting order, the document that tasks all servicemen and women to assume any new role, stated I was to take over the job of SO3 Research. Decoded, the 'S' revealed I would be performing a staff function. Conventionally, this rudely implied I would be desk-bound, pushing paper, providing 'organisational' and 'intellectual' support to a senior commander

– flying the proverbial desk. However, through a critical sleight of hand for me, all the operational roles in the Mission were designated as staff positions, granted, with paperwork to complete, but no different from any other operational post. The '3' revealed the position was the lowest-grade staff level, for captains. SO2s were majors and SO1s lieutenant colonels. The term 'Research' bore no correlation to the research work of the FRU but alluded to an operational intelligence role. Reconstructed, as the SO3 Research, I was effectively the Mission's professional intelligence officer even though I was one of four Int Corps officers in the Mission: Spandau was run by an Int Corps SIGINTer, Martin was a full-time tourer and Russian interpreter and Guy was the assistant ground operations officer.

Neither we nor the half a dozen Int Corps NCOs in the Mission wore our own uniform, Cyprus green berets or beret badges. Most of us wore Royal Corps of Transport (RCT) stable belts, rank slides and cap badges stitched onto the ubiquitous navy-blue beret worn by a majority of Army personnel, except Martin who retained his Education Corps uniform as a transferee from them to us. We had RCT buttons stitched onto our formal No. 1 and No. 2 uniforms, and actually had RCT mess kits made and fitted. So the ruse was certainly not cost neutral. The rationale was to play up the liaison function of the Mission and play down the intelligence-gathering role. I don't believe the Soviets were fooled at all, certainly in terms of the officers, for a single minute. They would have had access to the Army List, which succinctly links names to regiments and corps. But what it did do every now and again, however, was to create the situation where real RCT officers would engage us in conversation, usually at social functions, and enquire about our transport background, puzzled that they had never heard our names before in any transport-related context. The only solution to end the farce was to explain the situation and make light of it. I sometimes wished I had not so meekly acquiesced to accepting RCT 'cover' and instead pushed for the Lifeguards, complete with metallic chest plate and plumed helmet, or at the very least the Black Watch with kilt and dirk. I suppose, more logically, I could have used the precedent of my infantry attachment to 're-role' as a Royal Scots officer but frankly that association, after my infantry attachment, did not appeal.

Those Army officers who had learned Russian to interpreter standard were the full-time tourers. Paradoxically, they were generally not professional intelligence officers. I think I am right in saying that very few RAF full-time tourers were linguists as the vast majority of them were aircrew and could not be spared for the eighteen-month Russian course.*

Frankly, I had little interest, and even less aptitude, in learning languages and would certainly have baulked at the complexity and novelty of learning Russian. I have always felt the draw of learning Arabic but when that chance came I chose to leave the Army instead – bad timing after a career founded so consistently on being in the right place at the right time, but I do not have any regrets. During my service I had learned German on my first long university break, and I did learn a bit of Serbo-Croat one-to-one with an instructor for two intense weeks prior to deploying to Bosnia. I did undertake a modern Greek course in Cyprus when I commanded the HUMINT unit there, but I have to say they were not the fondest of memories.

In the language of the Missions this meant that they, the full-timers, were permanently 'on pass' – the holders of the eleven passes detailed in the Robertson-Malinin Agreement, less the one pass permanently allotted to the chief of Mission. The practical implication of this was that they had permanent free access into the DDR via our entry point over the 'Bridge of Spies' at Glienicke, into Potsdam. We part-time tourers came on pass irregularly several times per year when the full-timers were on a break for leave, courses or whatever. Our prime responsibilities within the unit were to run the specialist sections that supported the collection and reporting effort.

Therefore, in reality, taking BRIXMIS as the example of Allied Mission effort, there were a number of offices, accommodating the seventy or so Mission personnel at the time of its demise, running along the corridor on the top floor of the block housing the BRIXMIS presence in the stadium

---

* The British preference for the 'jolly amateur' was not shared by the Americans. The bulk of their tour officers were intelligence professionals, trained and managed by the DIA as part of their Foreign Area Officer (FAO) programme whereby they were immersed into the issues and language of a region of the globe and deployable on any mission, from attaché service across the spectrum to touring as a Soviet FAO in the Mission. Probably an approach only achievable due to the economies of scale of the US military, but in my opinion it is the way to approach issues if you seek effective global reach. Within the Russian speciality, a tour in the US Mission was the most sought-after prize, and competition was consequently fierce. The British approach was to identify and select likely officers, then provide them with the training they required, including the Russian language interpreters course for the full-time tourers. What the French approach was, I have no understanding, but I would be surprised if at least a few were not co-optees of their foreign intelligence service, the DGSE.

complex. The Chief's office was at the far end nearest the Olympic Stadium with the adjutant, the Deputy Chief and the SO1 adjacent. On the other side of the Chief's office was the admin hive of the unit, crunching all the numbers and letters that kept the Mission breathing. 'Operations', sitting mid-way along the corridor, with a ground and air component, tasked the tours and coordinated operational liaison with our French and American allies. In doing this they ensured there was a presence within each of the three touring zones into which the country was divided, plus an additional ground, but no air, tour active in the 'local' area, with each Mission providing the local tour changing daily or every two or three days.* 'Weapons' was a small team and the oracle on all things equipment and technically related. The weapons officer liaised with the various technical staffs back in London and elsewhere. Their office was Chief-side of the ops room. Spandau historically had been responsible for providing the British interpreters for the one remaining prisoner in Spandau Prison. When Rudolf Hess died in 1987, the office reverted full time to its prime mission, processing the material recovered by the Mission's most sensitive collection operation, OP TOMAHAWK. I will elaborate later on this fascinating and unpleasant element of the collection capability. 'Geo', a two-man show, managed the mapping side of the unit's operational role. The DDR was roughly the size of Ireland and each tour could call on 1:50000 map coverage organised in books to cover the whole country. On occasion, tours would be tasked on geo-related missions to reconfirm or update something out on the ground – for example, the span of a strategically located bridge or some other feature.† All photographic support work was undertaken by the

* Area A was in the north of the country, B in the south-west and C in the south-east. Within each area, two of the Missions would mount tours simultaneously, an air and a ground. The 'local area' related to the relatively small, compact area with its disproportionately heavy military population abutting West Berlin to the west. Rotation would take place every three weeks with a geographic move clockwise. Perhaps it won't come as a surprise to note that the French would not always observe the protocol if they believed that in so doing they would miss the opportunity to capture significant intelligence. This reinforces my overriding impression that they ruthlessly pursued a France-first, non-NATO agenda and would suspend tri-mission cooperation at the drop of a string of onions in pursuit of it.

† One thing I clearly remember, but never understood, were the US-produced maps that we used. What are some of the most prominent and useful landmarks and aids to navigation? I would say churches – most larger villages have one, and even small towns may have several, maybe one with a tower and maybe the other with a spire, visible from miles away. But, perhaps in a prescient nod to wokism, the American cartographers had chosen to remove all churches from their mapping. Bizarre!

RAF-staffed 'Special Section'. They were the unsung heroes who developed thousands and thousands of rolls of film per year, produced stills photos and, of course, managed and maintained the Mission's extensive camera and lens arsenal – they equated to a busy, large civilian photographic shop. To illustrate not just Special's contribution but the increase in intelligence collection by the Mission as a whole, in 1963 a total of 1,100 rolls of exposed film were processed, from which 63,000 individual frames were printed. By 1984 the figures had risen dramatically to 7,324 and 343,386 respectively, an approximate six-fold increase. Special anchored down the corridor at its far end, with Spandau next to them – keep those TOMAHAWK-related odours and other unpleasant effects as far away from the leadership as possible. Then, of course, there were the tour officers' and tour NCOs' offices. Each routine tour consisted of an officer and NCO, plus driver. The French and Americans made do with two-man teams, but this approach, necessitated by the fewer number of passes they had originally negotiated in their respective establishing 'treaties', was not as efficient operationally. The drivers formed their own section, joint Army and RAF, and to a man were both exceptional tradesmen and good guys to have in one's team – most were stoic, humorous and resilient characters. Lastly, was my team, the 'Research' office, located on the other side of Operations from Weapons with an interconnecting door – nothing to do, as I said earlier, with the research work of the FRU, at least when I arrived on the scene, but that was to change.

The Research office was effectively the mission intelligence cell and was the single authority on the order of battle and the organisation and deployment of the Soviet and East German ground forces within the DDR. Its everyday currencies were the vehicle registration numbers and their logging and recording that identified which unit from which formation had been sighted by the tours. Tours religiously recorded these details for the Research office to process in order for unit movements to be tracked. A critical event in Mission history had been the deciphering, Rosetta Stone-like, of the VRN system used by the Soviets, which led to this clear understanding and accurate subordination. The comprehensive database storing all our information regarding the main target was called ELECTRIC LIGHT. It was maintained by a civilian contractor from Serco. We also maintained a much smaller database of *Stasi* surveillance cars and their vehicle registrations – not that this brought any great extra operational focus to things as, frankly, the *Stasi* mobile surveillance assets were about as easy to spot as a flying T-80 tank would have been.

Sinclair, the senior NCO on the three-man team – later to be replaced by John, who would remain in the office right to the very end – had been on the course with me, and his grip and drive were admirably augmented by Chris, the JNCO – I would meet Chris again during the Iraq War when, as a WO1, he worked for me on the HUMINT side in Basra Airport.

I must raise my right hand and admit for the record, from the outset, that I had very little interest in the minutiae of Warsaw Pact forces. I was never what the Intelligence Corps would have labelled an operational intelligence 'tradesman'. They were the specialists who excelled in supporting the conduct of conventional war fighting from the vantage point of their map boards and IT terminals, sitting in formation HQs. They were the craftsmen who assessed the enemy's strengths and weaknesses, and advised the commander on their likely intent in order to inform his battle planning.

If it does not sound paradoxical or contradictory, I was very interested in GSFG/WGF in a broader sense as our enemy, but I had little interest in the fine detail of where the various units within this giant military beast were barracked, how they were constructed and constituted as units and, although this was the office's main supporting tool for the tourers, what VRNs and two- or three-digit side numbers, painted on vehicles' turrets and bodies, identified the various formations out on the road and exercise areas.* This was detail that did not excite me. I think I have always been more drawn to larger-scale and wider and broader processes, procedures and systems and their application, rather than the detail and specific data that feeds and populates them. I guess this is why I have always hankered for, and been attracted to, the operational world. It certainly respects system, and it certainly demands one to be systematic, but favours larger-handed, bigger-picture approaches to achieve the desired end state.

It did not take me very long to understand the Mission's approach to getting the job done – it should not have, after having attended the BRIXMIS course, and I had had plenty of time to think things through mentally, preparing myself for the job. Very quickly, I became familiar with the

* As a minor, pedantic detail, NVA equipment was invariably marked and differentiated by four-digit side numbers.

methodology, tactics and the systems that underpinned BRIXMIS's unique position in the intelligence-gathering firmament.

I had absolutely no qualms about our role in providing a steady update of technical intelligence on air and ground targets, nor the value we added reporting on tactics and doctrine. I was extremely impressed with what I saw of Spandau's reporting. And I clearly saw, and appreciated, where all three Missions had contributed to the maintenance of the peace, monitoring both Soviet initiatives and their reaction to the international fits and starts that constituted the Cold War. It was clear to me how BRIXMIS had, in its own small but focussed way, played a hugely significant role in managing vital communication as the post-Second World War 'founding fathers' had originally intended.

But, this said, I did begin to rapidly develop some doubts as to how effectively the Missions would be able to react to the incontrovertible evidence that the Soviets were making their final preparations to attack the West from a standing start. Perhaps, more esoterically, I began to harbour questions concerning the extent to which tours bravely battled to collect even random and low-priority detail in order to identify and subordinate clearly unimportant moves of isolated columns and rail packets of equipment. Subordinate to the strategic-level monitoring mission, I was not sure, at this early stage of my exposure to the operation, if our exhaustive and exhausting approach to collecting information on routine isolated troop and equipment moves was sound or even necessary. Being candid, I won't say that my nascent conclusions were totally negative or completely jaundiced ones, nor fully formed or strongly held. I was just metaphorically thinking aloud and beginning to formulate some questions that courted some nagging doubts and uncertainties. I did wonder whether anyone else had, in generations past, or was, somewhere else along the corridor, entertaining similar doubts. Sometimes it is more prudent to just tow the established line. The Army is not renowned for warmly embracing its free thinkers, rebels or out-and-out heretics. My ambivalent formulations were certainly ones that, at the time, I kept very much to myself as a new boy, although I am sure those who got to know me well probably identified some cynicism on my part towards the pure 'research' element of the job.

Referring to a 'standing start', mean that, in the unlikely but still feasible scenario, the Politburo had made the decision to launch a surprise pre-emptive assault, based on an assessment that they themselves were either in a sudden and critical position of weakness or, alternatively, one of overwhelming short-term superiority, this would, for whatever real or perceived reason, have been precipitated without any prior warning or escalation of rhetoric or action – more than likely supported by the complexity and persuasion of a disinformation campaign that the Soviets were such masters at constructing and executing. Their pounce would have been as sudden, violent and deadly as the strike of a leopard or, (perhaps metaphorically more apt), a great brown bear.

Relative to my second reservation, I pondered if our approach to feverishly logging and recording limited movements of equipment by road and rail bordered on recording information for the sake of recording it. Had it become a routine, reflexive ritual? I know I might be overstating the case a little to make a point, and that that point has now been rendered irrelevant by the passage of over thirty years, but it seemed to me that we trod very close to, if not were following, the mantra that we had been doing it this way for years so we would and should continue to do it this way for years to come. The events of that explosive, revolutionary November relegated this musing and questioning to idle speculation. But I have wondered whether, if the year had ended as had the year before and the one before that, I would have openly started to voice these queries and doubts.

On the wider issue of being a vital and precision-tooled link in NATO's tripwire, in essence, as the decade wound down, no one really believed, if we exclude the miscommunication and fear of 1983 generated by Exercise Able Archer and the shooting down of Korean Air Lines Flight 007, that the Soviet Union would invade the West in the second half of the Cold War. I know the former Chief of Mission, whom Brigadier Jackson succeeded, certainly did not. The USSR was still very much perceived as our enemy, and we still had few doubts that we were locked in an existential ideological struggle that might last for generations yet; against this outlook, neither side could let down its guard and show weakness for a breath, but war on the north German plains seemed a highly unlikely scenario – such was the balance of nuclear terror. The consequences were too terrible to contemplate both in terms of damage to state and people and, especially from the Soviet perspective, to the survival of the socialist experiment.

I speculated reflectively, sitting in my chair in the Research office, that if Moscow did take the decision to launch westwards, the tours of all three Missions, active in all three zones and including the local area, would have been caught up in a very rapid sweeping, suffocating and monstrous movement of troops and weaponry. This would have lasted days, certainly not weeks. This mobilisation would have exceeded any prior movement witnessed in Mission history and would have steered crews inevitably to only one despairing and desperate conclusion. No matter where in the DDR a tour had been operating, it would have been clear the level of increased activity signified only one thing – an attack. The reality of the tripwire snapping would have been incontrovertible. Against this volume and intensity of activity, tours would have been unable to log all these break-outs from barracks to the emergency deployment areas (EDAs) there would have been just too much to take in. The issue would have been, without communications equipment – neither Allied Mission tour vehicles nor tour members were equipped with any communications equipment – relaying the enormity of the situation back to West Berlin in a timely and effective enough manner to provide any early warning. Tours would almost certainly have faced deliberate and aggressive Soviet counter-intelligence activity likely to have rapidly involved their very swift and systematic deaths. Soviet planning would, without any doubt, have vectored in these aggressive plans to neuter the Missions. Avoiding detection would have been a near impossible feat of escape and evasion for tour crews. Were they able to remain free for any length of time, it is unlikely that East German domestic phone lines into the Mission Houses would have been left to work normally, and the option to telephone directly into the West did not, of course, exist.

Back to my initial concern, distilled down through the lens with which BRIXMIS viewed the world, I entertained no qualms whatsoever that its mission to 'patrol' the choke points, the major road junctions and rail routes was valid in providing the confidence that the Soviet forces constituting GSFG/WGF were not readying or deploying to mount a pre-emptive assault across the IGB. This was the time-honoured and unchanging importance of negative intelligence. But in the absence of warnings of increasing tension from high-level political indicators, or tactical tip-offs from strategic intelligence sources, I did have concerns that there was any requirement to do more than sit passively watching and monitoring. The ongoing status quo, set against our existing comprehensive knowledge and understanding of GSFG/WGF's orbat, surely just required BRIXMIS to ensure the tripwire remained taut and in place. Perhaps, frankly, the furious 'running', as we drove past them, of individual columns of trucks or other equipment, not to mention the pressured, frenetic recording of every last APC and Ural truck on a rail flatbed during a rail watch, was unnecessary against this practical backdrop and strategic reality? Additionally did we need to push and strain to record the side number of a tank or an APC sitting inside its barracks? We knew who and what was supposed to be sitting inside each and every barracks. We would know soon enough, in due course, when any slight adjustment or change was made to this situation. Too much reflex, too little reflection?

I am now very aware that I am about to step on potentially sensitive and emotive ground. I am not categorically voicing or suggesting any criticism of Nick Nicholson whatsoever; I know from my friends within the US Mission, and indeed contemporaries of his within BRIXMIS, like Guy in the Operations suite, that he was a highly respected and competent professional. But did he really need to get a look inside that shed to see a vehicle that we knew could either be there, or would be in some other part of the garrison area? He knew and we knew this was the home of 21 Motor Rifle Division (MRD). Put brutally frankly, we knew everything about a GSFG motor rifle division and, by extension, 21 MRD held no secrets or surprises. If Nick had happened to catch sight of a new side number on a turret or hull inside that shed, would it have really mattered? It would have been one single new data point in ELECTRIC LIGHT. Did he need to push his luck that day in such a scenario, just as so many tourers had done before and after him? Wasn't he just uncritically following standing operating procedures (SOPs) because that was how he had been taught to tour, that is how his colleagues in his Mission and our Mission conducted themselves and that is what the Mission hierarchy, equally unquestioningly, expected and was the metric against which they assessed his annual performance? I have no answer.

So back to my tentative and troubling professional doubts: the continuing pursuit of routine tactical collection activity appeared to have gathered and maintained its own momentum. Tours devoted so much energy into mounting road and rail watches and recording these disparate movements of kit because that was what tours did and that is what tours had always done right from the very earliest days because that is what was undoubtedly required then as the Missions found their operational feet. From a purist's perspective, for an intelligence-gathering unit, this almost blind rush to crunch numbers had evolved into a situation to be avoided or, at the very least, regularly and rigorously re-examined, re-evaluated and re-sanctioned. The Mission had likely become, but I do stress only in this particular area of activity, the proverbial 'self-licking ice lolly'. Had we reached the undesirable state of being, in a way, in part, a self-tasking collection organisation functioning without sharply pointed and tailored tasking, or the challenging scrutiny, of the intelligence staff whose role it is to ensure that scarce resources are employed to gather intelligence in strict accordance to a centralised, coordinated collection plan that adds value to the collective enterprise? Was rigorous management of collection plans by the intelligence staffs further west required to refine this risk? I suspected the truth of the matter was that the decision had been made, dismissively and regularly rubber-stamped, that the Mission's tactical modus operandi was acceptable and unavoidable, the 'ballast' that complemented, and fell naturally and unavoidably out of, the Mission's main effort – maintaining that relentless watch that might signal the likely phased build-up preceding a first strike by the Warsaw Pact. The easy, unchallenging path of least resistance had been taken, retaken and ultimately institutionalised. The Mission was there, its status unique and independent of the national agencies, its command chain established, trusted and proven, and its funding chain guaranteed, so it was easier just to maintain this routine and accept the status quo regardless. If that was indeed the case, it was not a professional approach to directing operations. But all in the event, moot points, idle conjecture and specious argument, although perhaps worthy of debate and comment even if only academically.

But now, flipping the coin back over, and being totally pragmatic, had there been any other way realistically to task tours to do business? My answer: probably not.

Fortunately, Sinclair and Chris ate up the detail of order of battle, military number plates and other subordination tags for breakfast and really added value to the Mission's perception of its day-to-day collection priorities, regardless of my personal speculative analysis of its real intelligence worth. They were truly superb – I like to think I managed them effectively and efficiently and contributed positively, with plenty of humour, to fostering their high degree of motivation. Despite my sense that we were engaged in an activity a little too close to trainspotting – literally – I was realistic and certainly supremely grateful to be immersed holistically in the whole environment. The office added considerable value to the sightings that the tours logged and provided the perspective and context that confirmed, at the routine tactical level, both the presence of the normal and the absence of the abnormal. When, and if, significant moves of men and machinery had occurred and the detail landed on their desks, they would have been there to identify it, assess it and report it. Ultimately, my position in the Research office had brought me to BRIXMIS, another high-profile and hard-hitting collection unit, rather than some run-of-the-mill outfit somewhere else doing something even more tedious. It would have been ungrateful of me to rock any boats. Invariably it's the boat rocker who goes overboard first anyway.

On the personal, professional level, my time in the Research office gave me my SO3 staff ticket for promotion to major, which came to pass before I left Berlin. But beyond this, excitingly, and for me the most important factor by several leagues, it did serve as my key to opening the door to another civilisation – to look, touch and smell communism up close and personal in the Soviet Union's most valuable satellite. I considered myself even then to be fantastically privileged. Now, looking back with the perspective of hindsight, I consider the experience to be a unique and priceless adventure beyond measure that I had the supreme good fortune to be part of – what we all witnessed is no longer there in the shape and form we knew it. The belief system has gone, the culture has gone, the infrastructure as we observed it has gone, the country itself has gone and, of course, the DDR's sponsors and protectors, the Soviet Union and Red Army, have both gone. It's as though we had had access to the Inca Empire, the entity that occupied the Angkors of Wat and Thom, or some other fascinating civilisation from history – it was, without debate, the stuff of history books.

I also had at the back of my mind the idea that surely there was some mileage, some potential, to squeeze from the realities of our totally unique operational environment in terms of exploiting other opportunities – HUMINT-related opportunities. We were out and about, cheek by jowl, amongst our Cold War adversaries, every single day of the year. Surely, contact – 'point of contact' to use the human source handling terminology – as intimate as this could unlock intelligence of another sort and potentially initiate relationships for exploitation by other agencies, perhaps in the future, in other places? Although neither I nor anyone else at the time knew it, my ticket to the BRIXMIS party bought me the opportunity to witness one of the most fundamental changing events of modern history and, in operational terms, as a consequence, to push this approach, undertaking intelligence collection far more aggressively and over a broader bandwidth.

This said, life in the Mission definitely had a cocoon-like quality to it. We existed in something of a bubble, but as I will shortly illustrate, we did not, by any definition, live in a vacuum.

In terms of security, because we were familiar with the threat environment in which we daily operated, and followed vigorous, tailored and practical security SOPs, we felt confident and secure as individuals and protected as a unit, without any degree of paranoia or complacency. The Mission was a finely honed blade wielded by the professionals from whom it was constituted. We shielded ourselves from unnecessary overfamiliarity with the Berlin Brigade with whom we lodged, discreetly housed as we were on the top floor of the main building, a back fire escape allowing us to load and unload our tour vehicles, creating the minimum of fuss and attracting the minimum of interest from curious onlookers. The route we took over the Glienicke Bridge into Potsdam ensured our profile was minimised. Once 'over the Wall' into the DDR, we understood the security situation, as it affected the Mission House and followed common sense security protocols to negate threats posed by the locally employed staff who, we later confirmed, were being run as informants, and exercised discretion when using the phone link back to West Berlin, given the correct belief that all lines into and out of the house were monitored by the *Stasi*. On tour we were fully aware of the security environment in which we operated: *Stasi* surveillance; alert mission-aware Soviet and East German NVA sentries posted throughout the country; and state police and a population almost exclusively, without exception, loyal to the regime and likely to report all sightings of Mission vehicles and tour personnel. The degree to which we were able to control and influence our surroundings contributed to a first-rate defensive security posture. Underpinning this was the sure knowledge that the Russians were fully aware of exactly what we did out on the ground – as the cartoons I describe later in the book attest to.

What we were unable to influence, and certainly to control, was the wider backdrop of the gargantuan Cold War intelligence struggle that was being ruthlessly waged beyond our bubble in the rest of the world, not least in neighbouring West Germany. We consequently were blindsided and failed to appreciate at the time that, at least in part, SERB's vision concerning Mission activities came from what would then have been regarded as being a distinctly shocking source.

We were aware of the KGB and sister services recruiting agents in the West, although we felt insulated from this reality: it was an almost abstract threat and not one that we felt directly impinged upon us. But it now appears, however, that this view was completely ill-founded. The real reality was that mission security was totally compromised, the consequence of a functioning wider treachery, well beyond our own sphere.

It is now known that, due to a number of successful Soviet-run espionage cases involving US personnel – and it is not unreasonable to assume there may likely have been other cases within the NATO orbit that, for whatever reason, remain secret and have yet to be exposed – the Russians enjoyed absolute visibility of the bulk of intelligence collected by the US Mission throughout a significant part of the Military Liaison Mission (MLM) era.* As an extension, due to Allied intelligence sharing and the fusing of the combined product, BRIXMIS and the French Mission were also equally compromised. This included the existence of the operational crown jewel TOMAHAWK/SANDDUNE.

What does this mean?

Essentially, there is no other conclusion to be drawn than to have to admit that the intelligence collected in the DDR, on the ground – the analysis of all our sightings, all our technical scoops, all our understanding of Warsaw Pact war fighting doctrine – was significantly degraded and, in real terms, lost much of its value. In the wilderness of mirrors that characterised Cold War intelligence and counter-intelligence operations, the Soviets knew what we knew, and likely, the other side of the coin, by a logical process of extrapolation, what we did not know. Cascading down from this, they understood what our intelligence collection priorities – including the Allied Missions – must have been in order to remedy the gap in our understanding and the direction in which our policies and programmes were being directed. We must assume that, at least in part, they had had a good understanding of how comprehensive our appreciation of Soviet capabilities was and, particularly important, their technical prowess

* Aden Magee cites the Hall, Conrad and Trofimoff treacheries in *The Cold War Wilderness of Mirrors* as being instrumental in creating this disappointing state of reality. Hall had worked for the Soviets throughout most of the 1980s, Conrad from 1974 to 1988. The senior of the trio was US Army Colonel Trofimoff who had been recruited in 1969 and was not arrested until 2000.

as part of the unceasing qualitative arms race. This was especially important in that constant bilateral marathon to maintain potential advantages should hot conflict break out. Looking specifically at the Missions, it is also highly feasible that understanding where we were likely to deploy tours in order to fill those intelligence gaps meant the Soviets could direct assets, for example *Stasi* surveillance, more effectively to try and deter successful touring. Did this mean the blood, sweat and tears shed and expended by BRIXMIS was all for nothing? No, of course it does not. It could actually paradoxically be argued, in the reflected light of 20/20 hindsight, that levelling the playing field in this way may actually have helped preserve peace, but it does, to me as an ex-Mission member, take some of the shine off what, measured by any other metric, was stunning operational success. I feel we were all cheated and let down and we cannot help but carry this reality with us to the point at which we make the final analysis and weigh up and balance the contribution the AMLMs made to our victory in the Cold War.

Ultimately the Politburo, could have chosen to exploit the knowledge they gained through their ruthless espionage – perhaps with some potential implications for source protection *vis-à-vis* their agents providing the intelligence – to terminate the post-war treaties Malinin had signed with the three Allied Generals, but chose not to. This strategic decision could only have been based on their assessment of the relative value of their Missions operating in western Germany set against an acceptable level of damage wrought by the Potsdam-based Allied Missions – damage that could ultimately be largely controlled. It is now apparent that history attests to their conclusion that, in terms of hard *realpolitik*, they could tolerate the British, American and French presences because – as I briefly touch on in the penultimate chapter – of the gold mines in which their own teams toiled working out of Bünde, Frankfurt and Baden-Baden – what they produced was the orbat and technical intelligence we produced but with other wider operational nuggets.

7

Flashback

We see the world from where we stand, and are moulded and shaped by the sum of our combined experiences – all good clichés but no less apposite views of reality for that. It was turning back the clock a decade to my very first martial steps in uniform, to some of the places and experiences that played their part in shaping my military persona, that prepared me for the challenges of life in the Mission. These are the earliest influences on my career whilst still at university, on my summer breaks from study and my year in Northern Ireland with the infantry.

I had been very fortunate to be rated in the top twenty or so candidates who had passed successfully through the Regular Commissions Board (RCB) at Westbury in Wiltshire for the year of 1978. The RCB was the British Army's five-day intensive officer selection process. As such, I had been selected for my university Cadetship. The Army's perspective was that we had been 'talent-spotted' as a future generation of senior officers, so they were prepared to make a substantial investment to retain our services for as long as possible – this they did in terms of our salaries and their payment of our university fees.

My side of the commitment as a Cadetship officer, apart from successfully graduating, was to regularly attend the nearest Officer Training Corps on weekday training nights, travelling down from Keele to Birmingham to the main Territorial Army centre in the city in leafy Edgbaston. This actually turned out to be a welcome distraction from the sometimes claustrophobic existence of campus life. A more potentially intrusive and demanding requirement was to spend four weeks during the long summer vacations attached to an Int Corps unit.

These rapidly turned out to be positive experiences, not only providing some focus during all those weeks of summer lethargy, but importantly increasing my woefully inadequate hands-on knowledge of Army life and exposing me to some real-world military intelligence officers, soldiers and work-related challenges. The plan was then that I would return to Sandhurst to complete my post-university course before moving on to serve my twelve-month attachment to an infantry unit on operations in Northern Ireland, a little bit better rounded and grounded in my profession. And so it all came to pass.

My first summer foray saw me in the heart of Germany's industrial belt, the *Ruhrpott*, the string of towns blackening the industrial banks of the Ruhr river, at the Army Education Centre in Mülheim to learn German. My tour with the Mission very quickly vindicated this opportunity which, at the time, I was less than excited about. Being conversant in German, despite the thick accents in parts of the east, most notably the Saxon 'dialect', was certainly a useful adjunct to the Russian of the interpreters. In the break before starting my third and final year, the card I drew was much more promising – attached to the British forces in Cyprus. With my studies over and with a surfeit of time, in the Army's eyes if not mine, before returning to Sandhurst, I was attached to 3 Intelligence and Security Company in Berlin – another peach, and with some relevance to this tale.

So during my first four-week 'holiday', I learned German to an enhanced conversational standard, passing, with its military-slanted vocabulary, the colloquial exam at course's end. I found myself able to ask with confidence, but absolutely no hint of a Teutonic accent, whether the kind farmer would object to me parking my armoured bridge-layer in his farmyard next to the barn and pig pen. I wasn't a natural, and my German was not that good, but I could survive. On an early run to the pub, as I struggled to assimilate the long daily lists of vocab, I confused the barman by ordering an *Adler* rather than an *Alster*. I had requested an 'eagle' rather than a 'shandy', a situation quickly remedied but not without a dent to my confidence. I did improve, gradually but not massively, but never regretted the month's hard work. However, the course did serve as my first protracted exposure to everyday life in green. Being basically back at school, it was not exciting but I felt I fitted in; I enjoyed the company of my fellows and the cadence of military routine. It was a good experience.

Cyprus made a much more lasting impression and served to reassure me without question that this was the career path I wanted to follow. It was also where I deployed on my very first operation in the coastal city of Limassol, the nearest neighbour to the British garrison at Episkopi. Like a first girlfriend, this will always be a special memory.

I flew out to RAF Akrotiri in a Royal Air Force VC-10, Concorde apart, the world's fastest subsonic airliner. I was being attached to 11 Security Company, the Int Corps' security and counter-intelligence unit on the island. The company supported the British forces who, since Cypriot independence in 1960, had occupied the two sovereign base areas, or SBAs as they were commonly referred to, ceded to the UK as sovereign territory. The political reality was that service in Cyprus was constitutionally little different from a posting to the garrisons in Catterick or Bulford or, for the RAF families, Brize Norton or Lyneham. Of course, Catterick has no glistening Mediterranean Sea on which to windsurf, sail, ski or in which to immerse one's body parts. Nor does the sun shine there most of the year, and the food is much less varied and healthy. For those with a need for a little more adventure, the delights of Syria, Jordan, Lebanon, Egypt and Israel were almost literally a stone's throw away. For them, a flotilla of cruise ships also regularly left the port of Limassol, hopping from island to island along the archipelago of the eastern Med. Cyprus fully deserved its reputation as the modern military's ultimate 'sunshine tour'.

But the holiday brochure nature of life for the 10,000 or so military family members stationed on the island masked a very serious geopolitical and military reality. Cyprus's island location, moored off the then Soviet underbelly and buoyed against the Levantine littoral, assured its critical dual roles as a forward mounting base and the perfect platform for the collection of strategic-level intelligence.

The airbase at RAF Akrotiri served as the UK's springboard for any military activity 'east of Suez'. It is there resolutely waiting for the discord and disorder that have to be dealt with as the ebbs and flows of an ever-changing world alter the conflicting balances of international affairs. Its long runways also perfectly served the needs of the USAF U-2 flight based there. Its top secret missions over the region were top sensitive in nature and target but unavoidably public, as this amazing unsightly freak of an aircraft would regularly thunder down the runway and power away towards the stratosphere at a surreal rate of climb. Its return could be noted half a day later, mission accomplished.

Inward-looking and antisocially discouraging casual visitors, 9 Signal Regiment sat discreetly at the other end of the island on the Dhekelia SBA, in the 'lozenge' of Ayios Nikolaos. It has a grandstand view looking out over a no-man's land at the eerie emptiness of the deserted town, formerly christened Famagusta. When the Turkish military invaded the island in 1974, one of their strategic objectives was to take this once bustling holiday seaside town. Since then it has been cordoned off and lies deserted like a resting film set, renamed, in the vigour of victory, Gazimağusa – very sad.

Less strategic but a no less effective rivet in terms of the security of the British presence on the island, 11 Security Company was headquartered in Episkopi's Army garrison in the western SBA, with a forward detachment in the east. It existed and survived because Cyprus has always been home to one of the biggest Russian diplomatic presences outside of the major Western capitals. It has always been heavily suspected that a disproportionately high percentage of these 'diplomatic' staff serve as intelligence officers. Their role would be to use Cyprus as a benign and permissive base from which to mount foreign agent and other intelligence-related operations in the region and, of course, target the British military assets based there.

Beyond this, the island also served as a relatively safe meeting place and logistics centre of operations, despite the periodic appearance of suspected Mossad agents, for Arab terror groups – it would never be wrong, even pre-9/11, to do whatever could be done to maintain an awareness of their presence and activities wherever and whenever feasible.

When I arrived in Berlin in the summer of 1981 for my final four-week attachment, my university finals completed and my sights now set firmly on completing my post-university course at Sandhurst, the city had been divided for the previous twenty years with absolutely no indication that this artificial state would ever be resolved in my lifetime.

During my attachment, whilst on a 'wire tour' with the company, I was able to pry a piece of concrete from the western side, of course, of a quiet and unattended stretch of the Wall. It was a rare surviving length of old Mark 2 Wall. Mark 1 had been the original hastily strung wire and bricklayer's Wall; Mark 3 was the latest reinforced and 'permanent' structure. I did it as a conscious act of defiance, important in my eyes but essentially petty, against the opportunism and brutality of the communist occupiers, and I suppose in part to obtain a trophy to demonstrate my desire to act. I still treasure it as a unique piece of Cold War history – I also own a couple of pieces of post-November 1989-procured concrete, but these shards, if collectively reassembled, would dwarf the Great Wall of China, more numerable than the slivers of wood from the Cross. I at least know for certain that my piece of Mark 2 is genuine.

Berlin was unique. Nicosia, the Cyprus capital, is also a divided city since the 1974 Turkish invasion, but somehow its symbolism is of a lesser order and grandeur. Berlin was truly both an amazing 'real' city and an incredibly exciting place to be as a Cold War icon and psychological weapon. Both blocs regarded their cake slices as centrepieces in the non-stop propaganda war, but frankly it was a war that West Berlin won easily on every front, perhaps except, tangentially, in the battle of 'most reconstructed historic public buildings' – the communist regime had devoted considerable resources to beautifying their Berlin, rebuilding from the rubble of 1945.

The west was a twenty-four-hours-a-day, seven-days-a-week advert for Western capitalism. It throbbed with bars, restaurants, clubs, theatres, cinemas, swanky shops and boutiques and an unrivalled public transport system. In contrast, the east, from the numerous points along the Wall over which one could peek, was grey, decaying and demonstrably as uninviting as gruel. Helmut Kohl was the most correct in his whole political career when he once observed that crossing into eastern Germany was like travelling back in time forty years. As I was to see first hand, as a BRIXMIS-hand, he was absolutely 100 per cent on the money. (However, he was absolutely 100 per cent incorrect in his promises that the post-unification budget would not lead to tax increases.) Not everything was bad there, but the positives of the DDR are not a discussion for here.

West Berlin was further sub-divided into a French, American and, of course, British zone. There were no corresponding lines or walls, or even signs I seem to remember, to demarcate these boundaries, and the Allied presence did little or nothing to subvert the character of this half of the old capital.

Known as '3 I Spy' to the British military community in Berlin, 3 Int and Sy Coy was certainly one of the most interesting British units in the garrison, and also within the Int Corp's order of battle anywhere in the world. Its Berlin-centric role was primarily to cross through Checkpoint

Charlie into the east of the city to conduct reconnaissance-based 'flag tours', mirroring the role of the Mission on a much smaller and more geographically restricted basis. These tours would record Soviet military activities and monitor the state of their readiness, and the levels of activity around the Karlshorst headquarters, incidentally a location that Putin, as a KGB officer stationed in Dresden, would frequently visit from 1985 to 1990.

Overt Red Army intelligence-gathering flag tours, conducting reciprocal reconnaissance on our side of the Wall, as mandated by the strict reciprocity of the governing post-war agreements, would be kept under surveillance by 3 Company's small covert surveillance team. Exciting Boy's Own stuff but also working to very real operational priorities set against the contemporary mindset of the Cold War stand-off.

The parallel role of 3 I Spy was to provide overt protective security support to the command in order to sustain a security posture robust enough to deter and ultimately defeat Soviet penetration. Important strategic assets stationed in the British zone, in addition to the tanks and men deterring a more aggressive re-run of the 1948 airlift, included the SIGINT collection station – or 'The Hill' as it was referred to because it sat on the highest point in the city, a man-made 'lump' constructed from wartime rubble – the airfield at RAF Gatow and, of course, the Berlin HQ including the BRIXMIS offices. I had no contact with the Mission during my four weeks and actually do not even ever remember seeing a tour vehicle during my whole time. Equally, no one elaborated on BRIXMIS's role or mission in my presence – need to know. It was almost as though the unit did not exist and, as I later experienced personally, this was the way the Mission liked it.

I quickly settled into the team. I absolutely loved every minute of my initial brief, albeit limited, immersion into this operational field. Here I was, just 21, dropped onto the very pulsating heart of the Cold War espionage stage before I had even really started work. Great psychology on the part of the Corps, and Colonel Carter, the retired officer juggling with all officer postings within the Intelligence Corps in terms of proactively addressing potential job retention issues. So sitting here in the offices of the company's 2IC it was, with tremendous excitement, that I read my first files annotated top and bottom with their red-ink SECRET classification.

Disappointingly, pretty much every one of them was an anticlimax. What I quickly learned to understand is that there is, usually, absolutely no relationship between the security classification of an item of intelligence and its juiciness. A TOP SECRET report in a locked high-security cabinet may well be as mundane in the reading as the pile of files lying untidily on a desk with their lowly RESTRICTED marking. In this analogy, the TOP SECRET classification is used to protect the source of the intelligence. This means essentially from whence or from whom the intelligence came, and the means by which it was obtained, rather than the degree of titillation the text excites. A simple and flippant example: a foot patrol in Belfast is out on the streets in the heart of Republican west Belfast. It's raining and 2.00 a.m. The patrol members pass a parked car gently rocking on its suspension. The patrol commander finds his torch and cautiously peers in through a side window. A known IRA gunman is in a wholly compromising, and not to say illegal, position with his older, unmarried sister who has a learning disability. On returning to base, the patrol commander, still chuckling to himself, writes his brief patrol report and briefs the company intelligence sergeant. Meanwhile, let's say that in Berlin, the Signals squadron on the Hill has intercepted a random Russian communication. It is reporting that the commander of 123 Motor Rifle Regiment, marooned on the Polish border in the barren north-east corner of the DDR has also been caught, coincidentally, equally compromisingly with his sister in the back of his UAZ-469 jeep. Whereas the paper gem in Belfast will likely be stamped with a desultory, but enthusiastic, RESTRICTED, (meaning, after use, flush thoroughly and please do not forget to wash your hands), on the Teufelsberg, a bored clerk will probably disinterestedly stuff the flimsy page atop a file tray already stacked with other largely irrelevant reports all sporting imposing TOP SECRET protective markings. SIGINT is a highly valued, highly sensitive strategic asset. When I joined the Corps, its very existence as a collection tool was not even openly alluded to. A foot patrol in Belfast, in terms of its inherent value as an intelligence tool, is just a bunch of guys with guns out taking a walk. This is not to say, of course, that a patrol of observant professional soldiers may not notice something that turns out to have significantly disproportionate consequences.

The file I remember reading most vividly dealt with an East German army officer working with diplomatic cover from a mission abroad. I think his name, undoubtedly assumed, was Major Berger. He had been traced as an intelligence operative, probably working for either East German military intelligence or for the HVA, the foreign intelligence service. His passport-style photograph stared back at me from the report. This was the face of the enemy, the evil face of communism that I had declared myself ready and eager to fight not so long ago as a schoolboy. I felt equally chilled and thrilled at this closeness and false intimacy. Our paths did not knowingly cross during my four weeks then, nor as far as I know at any time later in his homeland. I wondered a number of times, once I had joined BRIXMIS, whether he was just over the Wall, stationed in the east of the city or perhaps in a unit or headquarters somewhere in the wider DDR, or even abroad again on another foreign posting. I always often wondered who he was as a man and what happened to him when the Wall came crashing down.

I was not yet able to cross into the east, nor obviously, as I was not yet trained, take part in surveillance runs. I did sit, however, on a number of occasions 'eyes down' just beyond the deliberately intimidating wire, barriers and armed guards of Checkpoint Charlie and 'triggered' Soviet flag tours crossing over into West Berlin in their dark green Russian-made military saloon cars. It was never confirmed to me, but like their cousins in SOXMIS, operating in western Germany proper, one of the three equivalents of BRIXMIS, they undoubtedly performed the odd HUMINT-related support task, given the ease with which they could access the west.

I spent the majority of my month blistered onto the security teams as they made 'home visits' to their 'patients', quickly starting to get a real feel for this great city. I was lucky enough to make a recce flight along the Wall in an Air Corps Gazelle helicopter and took part in a number of unofficial 'patrols' along sections of its length – absolutely fascinating to experience this 'monster' at first hand. The Wall's physicality and the psychology of its sinister intent were undeniably and monstrously impressive, but a hard-to-comprehend testament to man's stupidity and willing surrender to subjugation by an outdated, impractical, unworkable idea. On one of these patrols I had to laugh, looking at one of the East German *Grenztruppen* military Border Guard guard towers. It was explained to me by my corporal 'guide' that the odd-looking glass panel in one of these regular and evenly spaced towers was in fact an incorrectly fitted pane of one-way glass. It had been installed back to front – we could see into the tower; the guards could not see out. Well done, Mr Honecker.

My very positive experiences as a military visitor by day were matched by my excursions as a tourist. As guarantors of the city's continued freedom, servicemen enjoyed the gift of free travel on the superb network of buses, the underground *U-Bahn* and older *S-Bahn* trains. There was no excuse not to travel far and wide, taking in Berlin's lakes and forests – flying into Tegel Airport, one realises what a relatively small but incredibly green city it is – the various distinct districts, posh Grunewald; the Turkish quarters of Moabit and Kreuzberg; the US presence to the south of the city – with their fantastic PX military shopping complex – and the French in the north; the upmarket shops of the Ku'damm, including the famous KaDeWe department store; the partially destroyed Blue Church; the Reichstag, with its pre-ugly and incongruous Norman Foster glass dome; the Brandenburg Gate and the draw of what lay beyond if one could pass through the Wall standing sentry there; and the main viewing platform overlooking the no-man's land of legendary Potsdamer Platz, with the still visible mound of Hitler's *Führerbunker*, and on into the expansive vista of socialist greyness. It was easy to spend a whole day moving from point of interest to place of beauty without even being aware of your imprisonment within the concrete stockade. I even managed to take in a strictly Division 2 Hertha Berliner Sport-Club (BSC) home game in the notorious 1936 Olympic Stadium. It still boasted the spartan boldness of Nazi architecture, but tickets for the forces of 'occupation' were also free. Incredible.

The *S-Bahn* was actually owned and operated from the east whilst the much more modern *U-Bahn* graphically represented the superiority of western technology and public utility. Paradoxically, one of the *U-Bahn* lines passed under the Wall stopping at the eastern station of Alexanderplatz. Shots of the city invariably portray the tall needle of the TV tower topped by its silver globe, firmly anchored to the south-western edge of the square. Military personnel were ordered not to take this looping line and were warned about the periodic identity checks that the *VoPo* (The *Volkspolizei)* East German civil police conducted on trains halting there. Of course as a Berlin novice I jumped on the wrong train and with a sinking sensation noticed that the name of the upcoming station was indeed Alexanderplatz. All sorts of scenarios flashed through my mind, none of them with particularly uplifting conclusions, and a few of them featuring time in Soviet custody. In the end of course nothing untoward happened as I sat there grimly, fists clenched in anticipation of the modern era Gestapo pushing on into my carriage. Eventually after what seemed like half the morning, the train recommenced its journey to return to the subterranean west. Either this day the *VoPos* were taking it easy; they had arrested their monthly quota of western spies infiltrating via this backdoor; or (more likely) the whole story was a ruse to dissuade squaddies from mistakenly alighting here and creating issues as they attempted to force an exit out of the station at ground level.

Diurnal fun was equally matched by nocturnal entertainment. It is not incorrect to state with confidence that Berlin caters for tastes in all arenas of personal leisure activity. Its restaurants cover every known regional and national cuisine, and it is the home for every type of club, from comedy to upmarket knocking shop. One of the more fun and slightly less sordid venues was a large bar/restaurant where each of the numbered tables was equipped with an internal phone. The concept was that if you wished to gain an introduction to the occupants of any other table, you simply dialled them up. All very pre-internet and pre-speed dating. A more earthy venue featured a large bath, mounted, for all to see, up on the central stage to which guests were called forth to bathe themselves and their female hostesses. Good clean capitalist fun. No more on this subject.

As a tragic footnote to my short stay in the city, on my first tour with the FRU in Belfast I received a bizarre call from Justin, the OC of 12 Int & Sy Coy, the holding unit that provided administrative support to all the intelligence sections established in Northern Ireland and, at this juncture in time, also to the FRU. He made reference to a very sad incident that had recently occurred in faraway Berlin, an alien world to the alien world of the Province. In the depths of whatever black place he had found himself in, the senior warrant officer of 3 Int & Sy Coy, whom I had, of course, met and had some brief dealings with but not got to know well, had signed out one of the unit Browning pistols and shot himself in the head in the unit toilets.

Justin asked me to visit his estranged widow, living in her native Bangor, out on the coast east of Belfast, and deliver to her a two-fold message. Firstly, to inform her that, currently, her deceased husband's ashes had been misplaced but would be found. And secondly, as soon as they had been located, they would be en route to her in the post. I was frankly shocked at the Pythonesque nature of this tasking, and wondered how on earth I would be able to find the right words to mollify her dismay, distress and no doubt disgust, and ease her continuing sense of loss. If nothing else, the blend of character and talents that mark the British Army officer ensures our infinite adaptability and, having located her address in one of my operational map books, I drove across the city to deliver my messages.

I found her house on one of the innumerable suburban streets of the seaside town and parked up a couple of hundred yards down the quiet avenue in front of an identical red-bricked semi. As I opened the garden gate to cross to the front door, it promptly fell off in my hand. I hastily wedged it back into place, speculating on what other disasters awaited me.

In the event, things unfolded calmly and straightforwardly. She was a sympathetic and understanding woman who received my shocking, and frankly unacceptable, news with grace and composure. There was little more that a 24-year-old, with scant life experience, could say in the circumstances. I had not really known her husband nor, at that stage of my life, suffered any close personal loss myself. I quietly left, successfully negotiating the broken gate.

Looking to the future, the last time I have had to take on a related task, I was back at SIW in Ashford working as the second in command and training major. It was my turn, again, to cover as the weekly duty field officer. Late in the evening a signal came in that the husband of a TA student on a course in one of the training units within Templer Barracks had just died. It transpired the student was from Northern Ireland and her husband was a serving member of the RUC. The signal gave cause of death as a heart attack. As the duty field officer, it was my responsibility to inform her of her husband's death.

She was staying in the sergeants' mess as befitted her rank and I was quickly able to identify the room number. Because of the circumstances I did not require to be formally invited into the mess as a visitor. I knocked on her door, the time now close to 11.00 p.m. After a brief pause, the door opened to reveal a woman in her late 40s. She looked at me and I could tell she instantly knew my purpose. My heart went out to her as she visibly seemed to deflate in front of me as I witnessed the stunned expression on her face, co-existing in mortal combat with the swirl of emotions as the rush of change that now confronted her swept through her consciousness. It was not a pleasant ordeal – I have had to pass on the news of the death of two close family members, but these challenges somehow seemed easier and were less upsetting.

I settled and calmed her and she began to talk. As she processed the situation, she now seemed relieved to hear her husband had died of natural causes and explained that what they had feared for so many long years was the prospect of death at the hands of the terrorists. It seemed so perverted and warped that news of a heart attack could bring solace in a situation that most spouses anticipate with dread their whole married lives.

Despite my understanding and appreciation of intelligence work – my wider HUMINT grasp developed in the Province, pointedly my grasp of the subtleties of photography that I had learned on the long surveillance course, the ins and outs of effective anti- and counter-surveillance and my own personal catalogue of Soviet equipment, on the BRIXMIS course, all learned later – it was my infantry attachment, consolidating the basic tactical and fieldcraft lessons I had learned at Sandhurst, that best prepared me, laying the solid foundation that gave me the confidence for meeting some of the more practical challenges of touring in the DDR. Holistically, I felt I was supremely well prepared for my posting to the Mission.

I had hoped that, as 1981 advanced towards the new year, I was in sync to serve with the Black Watch on their tour of duty in Belfast. I am confident that memories of my twelve months would have been fonder had I worked with this Highland battalion. But I take the view that, especially with the passage of time, even bad experiences mellow into good. Looking back now, I am happy. There were good times and I certainly, without question, learned some useful lessons for the future having seen the Royal Scots up close and personal for that year. Deploying as a platoon commander, and then a make-do stand-in company 2IC, was positively challenging, and Ballykinler base, in the lyrics of the famous song 'Where the mountains of Mourne sweep down to the sea', was a great place to be based.

1 RS was in Northern Ireland on an eighteen-month 'residential' tour. This contrasted dramatically with the four-month OP BANNER emergency tours upon which most 'roulement' infantry units in the Province served. This increased continuity had a fundamental impact on the serving soldiers' quality of life. Foremost, it meant they served accompanied by wives and children, if they had them. Families lived comfortably and safely in secure housing 'behind the wire', to use internment terminology.

The four companies within the battalion structure would deploy to the battalion's TAOR, their 'patch', in the north of County Armagh, including, temporarily, the city every fourth month for four weeks at a time. During this period they would mirror the activities of their emergency tour cousins: patrolling, resting and taking on internal base-guarding and admin duties. When not deployed, they would sit, ready and waiting, within the confines of Ballykinler to be deployed, as the brigade reserve, to any developing hotspot that required a manpower surge. The Ballykinler residential battalion was part of 39 Infantry Brigade based in Lisburn and responsible for security in the east of the Province, mirroring 8 Infantry Brigade in Londonderry to the west.* During the rest of the cycle, time was spent on routine training to keep basic skills up to speed, on leave and freeing men to attend various professional courses.

Frankly, I found this phase of the cycle pretty tedious. It quickly became apparent to me that for most of an infantry officer's life, he would find himself waiting with undue expectation, despite the danger and discomfort, for the distraction of an operational deployment to break the monotony of life in barracks – I thought back to my early adolescent visions of a life with the Queen's Regiment as an infanteer and thanked the gods for their consideration. Being in barracks did, however, mean that, particularly in summer, families could enjoy being on the beach stretching outside the rear gate, to sail, windsurf and swim, if hardy enough. They could also take advantage of time in the Mourne Mountains, and exploring the string of coastal towns along the County Down coast, with due security consideration.

I quickly assumed command of my platoon from the outgoing OC, a decent young lieutenant. It was a shame he left on posting. My platoon sergeant was a small, wiry jock, atypical for the infantry. There was an intellect and air about him that marked him out. He had served on a previous tour as a unit source handler. In those days, there were members of the battalion intelligence cell who handled the day-to-day background, observation-based, low-level intelligence from informal meets with what were described as casual contacts (CASCONs). This background created a degree of mutual affinity and an immediate bond between us, as he was, of course, aware of my intelligence cap badge, even if I was brand spanking new, without an iota of field experience. I learned from him and he guided my first days and weeks. Unfortunately I was not to benefit from his safe and steady pair of hands on operations. Then, with the platoon divided into its two constituent 'multiples', we went on our separate patrol paths. His mentoring mantle fell to an older corporal, in my eyes fussy and distinctly

* Later, in 1988, a third brigade was re-established to assume sovereignty in the southern border areas, including the 'Bandit Country' of South Armagh.

lacking in grit and charisma. We did not hit it off particularly positively.

The soldiers themselves, the 'jocks', as all Scottish soldiers are warmly referred to, were great as a bunch, with some true individuals within their ranks. My greatest issue now was being able to follow their every spoken word or, more specifically, not being able to. It took me literally a couple of months to 'acclimatise'. I think they found my less rigid and less formal approach to them at first a bit unbalancing but they quickly adjusted and I found them, to a man, loyal, consistently amusing and positive in all adversity. On the ground, they were hard working and professional.

I did not have to wait long before leading my first-ever live operation. My real 'hard' operational career began with it and launched the professional roller coaster that led to my first time sitting in the back seat of a G-Wagen, camera box at my feet.

The prospect, and reality, was incredibly exciting. We were to deploy into South Armagh in support of a large brigade cordon operation covering the insertion of a Royal Engineer plant into Crossmaglen, to be used to further reinforce the already Fort Knox-like base. It was also to be my second flight in a helicopter and my first in a larger troop-carrying utility chopper – before the end of my career I was to make dozens and dozens more in every rotary airframe the British military operated, except Apache helicopter gunships. I have laughingly asserted a number of times that I'm sure I had more flying hours than many junior Air Corps pilots. Helicopter transport in South Armagh was routine and vital, except for undercover troops, as PIRA landmines had made the roads prohibitively unsafe.

The sensation of power and motion as the Wessex pulled a final tight over-bank, descending onto our hilltop drop-off point was something I will always be able to relive in my mind. As it hovered 3 or 4ft above the ground, the three four-man bricks inside moved as one in well-rehearsed and fluid order to the open door for the loadmaster to tap them out. When my turn came, I experienced an unconquerable burst of power and energy as I hit the ground and sprinted to take a prone position in the clock face of all-round defence. We were ready to return enemy fire to protect the retreating helicopter. This was the most real experience of my life to date. Here I was, rifle pressing uncompromisingly into my shoulder, experiencing the moving air of the wake of the departing Wessex on the back of my neck, as I scanned my twelve o'clock for the enemy. I felt ready to take on the PIRA single handed.

The next three days were spent dug in on the crest of our hill overlooking the main road running south to the infamous border town of Crossmaglen – XMG or simply Cross, as locals and security forces alike referred to it, probably the only single thing they collectively agreed upon. Cross was the closest permanent security force base to the Republic of Ireland. It housed a joint Army and police post to which patrols incessantly returned to relative safety and bomb-bursted out into potential danger through its front gate into the staunchly Republican streets and lanes of the surrounding countryside. More British soldiers had lost their lives in these few square miles than in any other part of the Province. This included Private Barucki, after whom the lone 'sangar', providing an unobstructed view onto the main town square, was named.

The IRA was a permanent force that was well respected by the Army. They had fully earned their reputation as champions of the Republican cause and the most professional and ruthless terrorist grouping, not just within the Republican terror network but anywhere in the world. They knew every field, every gate, every hedgerow, every re-entrant, every feature of this land and exploited it boldly, skilfully and with deadly malice. As British soldiers, we prided ourselves on serving tours here of four months or, in our case as members of the FRU, several years, perhaps returning for another tour in years to come. Senior members of the XMG ASU had been on active duty there for twenty-five years by the time the ceasefire came. If they weren't actively respected by all members of the community, the fear they generated guaranteed support and aid.

Thankfully the weather remained mainly dry for the operation and we used our bird's-eye view to ensure the PIRA was not able to mount any quick response operation against our convoy within our arc of observation. There were no significant peaks to dramatise the topography, but the rolling, rounded, rocky hills from Slieve Gullion to Camlough Mountain added an enduring and long-suffering drama of their own to the vista as they rolled down from the east to the flatter land to the west. The landscape in the gentle valleys was so green. Endless blackthorn hedges delineated the fields and boundaries of the small family farms as they had down through the centuries and generations. The land was exclusively given over to livestock farming. The only other economic activity was generated by the odd small haulage company and the black money of the thriving cross-border smuggling gangs. Local syndicates, all franchised by the PIRA, moved animals to and fro over the unmarked border with southern Ireland in an unceasing and crazy dance to extort EU subsidy money. A first glance decried the claim, but the reality was that there was money, and plenty of it, passing through the pockets of these connected operators.

We maintained a twenty-four hour sentry to ensure we were not surprised by an attack, using night-vision scopes during the hours of darkness. The first morning I remember washing briskly in the collected water of a puddle within the perimeter from the small amount of overnight rain in order to help conserve water – soldiers in the field never have enough, and more than most assiduously try not to take this most precious commodity for granted. Later we were visited by a local farmer who amazingly delivered some fresh bread and eggs. He reappeared the next day. This small gesture was as deeply touching to us as it was dangerous to him. Collaboration in any shape or form was not tolerated by the IRA, but it demonstrated that there still remained a streak of humanity and a desire for 'normality' in the hearts of some of the people of South Armagh – an interesting thought that I took away with me, sitting sandwiched on my canvas bench on the Wessex that dully powered in to take us home.

I quickly adapted to the cyclic rhythm of our deployment schedule. I led a number of three-day patrols from our forward base in the newly constructed Drumadd Barracks on the outskirts of Armagh city and occupied the forward bases in Middleton and Caledon, small border towns looking across into County Monaghan, from which we mounted local area patrols to dominate the area, demonstrate sovereignty and promote security.

Armagh is a staunchly nationalist city, that is home to two cathedrals and the women's prison. It also supported Gough Barracks, the home of the regional RUC Special Branch, co-located with the police interrogation centre, and the large regional police headquarters – on later deployments this would become my home. Standing in for the UDR, which had military operational responsibility within the city limits, a patrol we mounted in the backstreets of one of the hardcore estates elicited a barrage of rocks and bottles from over the roofs of the adjacent housing. Probably youngsters, nothing to be done, and no injuries. Small beer.

On the longer patrols we, as foot soldiers, would carry everything needed to maintain ourselves, whether in the cold and damp of winter or the mediocrity of summer. We would set up a patrol base, usually in a copse or on higher ground, in which to leave packs and personal equipment, and patrol out on localised forays to show the Crown's attendance, leaving a small security presence behind to secure the base. In this way, a greater swathe of our TAOR could be covered rather than limiting the flag flying to the immediate environs surrounding permanent fixed bases – all good counter-insurgency tactics.

Our aim was to provide that element of reassurance to local people who felt insecure whilst deterring those who supported the aims of extremism. This close proximity to the local population would also generate low-level intelligence and initiate low-level dialogues that could be developed by the 'spooks' further up the food chain, all the way up to the FRU. There was also another real, tacit and perhaps cynical objective. Our presence also served as the red rag to draw out the PIRA bull and entice them into open combat. In the vocabulary of 'grunt speak', the green army were 'duty targets'.

Routinely, or in the event we were discovered by local 'dickers' – usually kids tasked by the local 'godfathers' to report on troop movements around their homes and haunts – the patrol would move to occupy a new patrol base the second night and any subsequent nights. At operation's end we would be extracted by helicopter, or occasionally in laughingly designated 'covert vans'. These civilianised Bedford Transits – no company logos, blacked-out rear windows, somehow usually odd Trabi-like shades of colour – could be identified by a sighted granny from 300 yards for what they really were. The vans would pull up to a pre-briefed location and the multiple would pile in, kit and all, for the drive back to Drumadd – exciting stuff for young soldiers but certainly well short of covert, and not overly professional in terms of health and safety.

In the event, all my patrols passed off uneventfully, but the challenge of constantly making tactical decisions and navigating the patrol effectively from A to B to C whilst maintaining morale and ensuring the welfare of my young soldiers was sometimes stressful and always tiring.

To break up the routine we regularly deployed on 'Eagle VCPs'. These were great fun and certainly the least risky task, if you discount the inherent danger of military helicopter flight, but the greatest force multipliers in terms of force projection. The four-man Eagle patrol would pack into the rear of the small Scout utility helicopter (which would be replaced by the much more agile Lynx), sitting two-by-two on each side, doors removed and replaced by a canvas utility strap strung across the empty door frame, boots resting on the skids. The pilot would drop the team at the marked map location the patrol commander would have handed him as he climbed aboard on the helipad at take-off. The patrol would move onto the road from the adjacent field and stop and check all cars and their occupants passing through the improvised checkpoint and radio through details to the company ops room for an OP VENGEFUL check – this was the intelligence-owned database storing all vehicle ownership details. It functioned synchronically with a personality database. Their combined utility provided a vital foundation for operations Province-wide. Five or ten minutes later, depending on the volume of traffic, the patrol commander would call in the Scout for extraction. Four or five drop-offs and pick-ups would be conducted before 'return to base'. The randomness of the Eagle patrols represented an existential threat to PIRA when moving men, weapons or explosives, or indeed deploying to conduct an operation. Agents reporting that revealed the presence of a helicopter in the sky very often persuaded PIRA to abort a job. The eye in the sky was all seeing.

I will always remember the face of one of my guys – a Black Watch private soldier attached, like myself, to the battalion – as we were climbing back into the cramped cabin when I rather prematurely gave the thumbs up to the pilot to lift off before Private S had fully and firmly re-seated himself. Re-seat himself he did, but I think I spoiled his afternoon.

Every now and then the spice of a fast ball would pop up at short notice to be inserted into the patrol programme. A tip-off had been reported by a source, most probably low level and of unproven reliability, stating there was a current threat to the life of an off-duty part-time UDR man. The inference was that a PIRA team would make its way to the soldier's home, either by car to the front door or on foot from a drop-off point nearby, and either shoot their victim as he opened the door or force access to the premises and murder him inside, likely as he slept in his bed. I had been tasked to move into position to cover this UDR man's home and establish an ambush to kill the gunmen, if indeed they did strike that night.

Part-time UDR soldiers did not enjoy the security of protected garrison accommodation and very courageously lived within their communities, on too many occasions paying the ultimate price. In reality, had the intelligence been more solid, the joint police/Army Tasking and Coordination Group – or TCG, a well-used acronym in the Province – based in Gough Barracks would have passed the task to a close observation platoon or, in the event of a highly credible threat, to the RUC Headquarters Mobile Support Unit or SAS. The three regional TCGs – one each in Belfast, Londonderry and one for the sticks – were a vitally important initiative providing a fusion centre in which intelligence reports from all sources, human and technical, could be merged, assessed and used in key operational tasking to exploit it. They were also responsible for imposing and managing operational out-of-bounds (OOB) areas. An OOB area would be requested to sanitise the location of a covert operation, mounted either by 14 Coy or the SAS, but not routinely by the FRU. The area-placed OOB could be very localised or cover a number of grid squares. Once imposed, the OOB area gave confidence to the covert operators that no overt green army unit, or routine police patrol, would be operating in their operational area and thus preclude the chance of a blue-on-blue gunfight – covert troops and overt troops do not happily mix. Big boys' thinking informing big boys' rules.

I made my plan. I identified the two small cut-off teams to secure the lane leading to, and running past, the isolated country bungalow. I designated a fireteam to cover the front and one to act if the would-be killers ingressed across country. As I worked through my orders sequence, I reached the critical paragraphs dealing with 'actions on'. These would be my team's orders covering all the potentialities that I assessed as more likely to occur and the contingencies of the 'what ifs' that could suddenly and unexpectedly crowd the decision-making process. My aim was to draw together all the factors that would ensure we killed or captured the terrorists, and saved our Army colleague's life.

All went to plan as we were dropped off by van in the darkness of the moonless late evening. All went to plan as we made our way to, and occupied, our pre-briefed positions. It was dry, not too cold and the potential import and drama of the occasion certainly kept me alert and poised. All went to plan as we remained concealed, silently, eyes sweeping our operational arcs of fire. We watched, we waited, we waited and we watched. All quiet. All went to plan as we folded the ambush just before dawn, with one minor discrepancy. No killers. It was unlikely that our presence forced them to abort. We had moved into position tactically and professionally, and it was highly unlikely that our presence was known, even to the residents of the cottage. More likely, it was just the all too frequent reality of inaccurate intelligence. All parties would live to fight another dark night.

A fast ball that I found affected me deeply was the kidnapping of another off-duty part-time UDR soldier. Thomas Cochrane was 55 and served with the County Armagh battalion. In essence, he was elderly and harmless, randomly targeted by PIRA because he felt a sense of loyalty to his community and to his country and was courageous enough to do his bit. I could not stop thinking about what the poor creature must have been going through. A massive police and Army search operation was initiated in an attempt to locate and free him.

Rarely, the whole of A Company, my parent sub-unit, deployed into the South Armagh countryside under the direct command of the OC, with me, as the ops offr, coordinating things on the ground at his side. We spent several days patrolling from potential hiding place to potential hiding place but, in the absence of both luck and hard intelligence, in vain. The IRA subsequently announced they had murdered him in retaliation for his part in 'atrocities' against the nationalist population – such bravery on their part.

I later heard that, in fact, Mr Cochrane had died of a heart attack – it probably made absolutely no difference to the grief of his family but somehow it made me feel better. Yet another utterly pointless and wasteful loss of life.

Halfway through my time in Ballykinler, just before the Cochrane kidnapping, the company's 2IC was posted and not immediately replaced. I moved from being a platoon commander and assumed his position for the remainder of my time with the battalion. This brought no extra spice to life in barracks but an entirely new dimension to deployments. The ops offr spent the month away deployed on operations in the Armagh city police station from where he directed the company's multiples out on the ground and liaised with the RUC, manning its control room co-joined by a dining room-type hatch to his ops room. This was my initial immersion into the wider world of operational command and control, a step up from the tactical to the operational. And it provided me with an invaluable insight into the police as an organisation and into the mindset of its courageous individual members.

Christmas and the New Year's festivities passed in peace until the bomb explosion in the small border town of Middletown. It had been built into a stolen car and driven across the border from the Republic through one of the numerous unmanned border crossing points. It was parked on the village's main street to target the certainty of one of our patrols passing it on leaving, or returning to, the small forward base. The PIRA driver would have locked the car and unobtrusively sauntered away to be picked up by a second car that had probably 'scouted' his route into Middletown. He would then have been whisked back to the safety of County Monaghan. A small PIRA team would have remained, concealed in the pre-recced firing point, with eyes down on the car bomb, waiting to detonate it remotely.

Around mid-morning their patience was rewarded: they would have observed their target moving tactically as the jocks patrolled towards the parked vehicle. The initial contact report from the multiple commander on the ground erupted over 'the air', shattering the languid routine of another day in the ops room. It hit us with the pulsing intensity of a super-charged lightning bolt. 'Contact, wait out' is the message all soldiers simultaneously dread, but at the same time instantly embrace, energised by the prospect of immediate direct action. All other callsigns on our battalion radio net suspended transmission, clearing the airwaves so my ops team could manage the situation and receive all further incoming SITREPs from the targeted patrol.

Miraculously the bomb had only partially detonated, almost certainly divine intervention rather than the hand of a covert operator, at some point in a targeted operation, tampering with the device in order to disrupt it. The vast majority of the car bomb's load had failed to explode. That which had gone off had caused some minor injuries to the nearest patrol members but nothing more grave. The patrol commander, a young junior corporal, coordinated immediate first aid and set about establishing a cordon to isolate the burning car and its contents.

The inherent complexity of such a situation is that immediate follow-up action has to be taken whilst simultaneously assessing the urgent practicalities of this against the contingency that there could be additional explosive devices set to catch reacting agencies. In the ops room we started to work through our checklist of post-IED actions. Casualties were evacuated by helicopter; ATO was tasked and routed into the location; supporting troops were ordered to reinforce the cordon; and other patrols, already out on the ground elsewhere in the TAOR, were warned of the heightened alert state and energised to be on the lookout for suspicious activity. Tick, tick, tick, tick – we marked off all the boxes and thought through and around the situation to identify any additional action we could take or coordinate. The RUC control room monitored the situation and played their part in the follow-up.

In the hierarchy of Northern Ireland atrocities, this attack barely rippled. But we had been very lucky. The reality could just so easily have been five or ten empty bunks back in Ballykinler and a handful of grieving families on the plane back to Edinburgh. From the professional perspective, it was a tremendous learning experience, and I was pleased that as a team we had not been found wanting and had performed our part well in handling the immediate aftermath. It marked the end of my baptism of fire that, in my own perspective at least, had moulded, shaped and hardened me into a real proto-soldier and officer.

Not all my experiences attached to the infantry were, however, about soldiering – a couple of my favourite stories marked the other side of life in the Army.

Talking about my fellow officers in the battalion, at major level, as company commanders, they were a diverse and mixed bag in terms of experience and personalities, but a decent and largely welcoming group. Most were married, accompanied by wives and invariably children, so were infrequent visitors to the officers' mess after working hours.

The single subalterns, the junior officers, were a different bunch. One had been at Sandhurst with me – he was a good guy, the son of a general. Most of the other ten or so were OK too, but there was a Macbeth-like threesome of obnoxious characters. They set the tone for the year when we four would meet again in barracks.

The Ballykinler officers' mess was something of a transit mess as well as the regimental home of the Royal Scots officers. In the eyes of both short- and longer-term visitors to the mess, my trio must have left a very poor impression on their regiment. They clearly resented outsiders in their 'home' and were not shy in passing on this impression. It tickled me to overhear them describing me on one occasion as 'that English bastard'. Paradoxically, I probably had a stronger genetic and cultural link to Scotland than any of these Surrey Highlanders, or more aptly Lowlanders, as they were. I tried to minimise contact with them, spending most of my free time with the Royal Marines officer who commanded the anti-gun smuggling detachment operating from the small minesweeper patrolling Carlingford Lough and an extended stretch of Ulster coastline. He was much more my idea of a professional military colleague and friend.

One of this obnoxious trio was particularly fat with well-developed porcine features to match. It always amused me watching him waddle about in full kit, not least because he was one of the COP commanders serving in the battalion's close observation company. How he squeezed discreetly into the tight, restricted confines of a covert rural OP – their role was to mount low-level surveillance – was worthy of speculation. I, for one, did speculate regularly.

On one specific occasion I was forced to 'break cover' and engage, specifically, with my fat friend. I was using the mess telephone, located in its own discreet cubicle on the ground floor, speaking to my girlfriend back in England. My *ami* tapped on the door and opened it, thrusting his flabby features forward stating that he needed to make a call as he had shortly to be off on deployment, no doubt squeezing into somewhere new until the position was compromised. I very politely informed him I would not be too long, but if he was in a tearing rush, there was, of course, the telephone located several hundred metres away, close to the main gate. He closed the door and I assumed he had rolled off to make his call. How wrong I was.

Within the shortest space of time, a fire hose appeared through the slightly open window illuminating the phone cubicle and, simultaneously, a jet of water began squirting inside. Of course before I could grab the end of the hose, I had become unseasonably wet, even in a country where it is not uncommon to be regularly soaked through. Needless to say, I precipitously ended my call and left the booth, with vigour, in search of this part-time fireman. I very quickly found him upstairs, hiding in the TV room by himself. I equally quickly suspended any of the reserve and self-restraint that I usually sported around the mess and flicked the 'On/Off' violence switch to 'On'. I came very, very close to exhibiting the most abject ungentlemanly conduct and the unleashing of a couple of jabs and crosses towards his wobbling jowls. Restraining myself with good judgement, and ultimately exercising proportionality, I explained to Fireman Sam that, if he ever pulled another stunt like that, I would, basically, punch his pseudo-jock lights out. In days of yore, incidents like these were the meat of duels. I would surely have put down the challenge in the sure confidence of victory, but alas this was 1982 not 1782.

That appeared to be the end of the matter. The next day, I signed out from the main gate and began my drive home on a spot of leave. My journey was progressing well. I was travelling back to England and had taken the Larne ferry and was driving through mid-Galloway, not a million miles from Lockerbie, when I began to hear and feel a juddering vibration from my front off-side wheel. I pulled over immediately. Upon examining the wheel, there was a rather drastic issue with the wheel nuts. I found that not one but all of them were loose and hence the rim was moving longitudinally against the restraining bolts. Now, I have never been a proponent of the concept of coincidence at the best of times. One loosened nut I could have accepted and blamed myself for not checking the car before setting off. But each and every one on the same wheel seemed to me to stretch any degree of coincidence to the point of incredulity. Tightening them up, seething, I decided I would let the matter drop but it was, and continues to be, obvious to me that someone had systematically used a wheel brace to knowingly attempt to cause a road traffic accident, with me playing the starring role. Three guesses – conceivably not fatty, who may or may not have been away by then, but certainly his henchmen. Any RTA can, of course, not only result in serious injury or worse to the driver and any passengers but equally also affect the well-being of pedestrians or other road users to the same degree. Now I sorely regretted not having well and truly punched the obese knacker.

When I met the commanding officer for my farewell interview, and to receive his appraisal of my performance following my annual confidential report, he commented that he realised things in the mess had not been easy for me and that I had not received the welcome I should have been accorded. I replied that really there had been no issues from my perspective and that I had enjoyed my time. A few of his petty, small junior officers might have had a go at drowning and killing me but it would take more than that to spoil things for me.

The officers' mess, as I've hinted at, was the epicentre of battalion life and, in accord with this reality, we would regularly attend formal black tie evenings. Every now and then, to mark some occasion in the regimental calendar, an even more formal event would be staged when the wearing of full mess dress was mandated. The best formal evenings involved the addition of wives and girlfriends. The atmosphere was much more conducive to having fun and witnessed far better behaviour by the young officers of Pontius Pilate's personal guard – yet another derogatory nickname for the First of Foot also known as the Royal Scots.

The most notable bad form I have ever witnessed at a military dinner occurred during the 'dining-in' of a couple of new officers joining the battalion, luckily minus wives. They had been force-fed drink after drink in the bar in the prelude to dinner. By the time we sat down, they were both in a pretty bad way. I was sitting obliquely opposite the most badly hit newbie, actually a decent guy. As another course was laid in front of him by Corporal T, the senior mess steward, the poor young lad's face, in sublime slow motion, lost all of its already pretty strained pallor, his head moved forward at the neck like a reaching giraffe, and a *tsunami* of partially digested, semi-curdled dinner, surfing on a liquid blend of stomach acid, beer, wine, whisky and Lord knows whatever other beverage, splashed across the table. There were several casualties, collateral damage caught up in this outrage, including the shirt and mess tunic of the athlete in question, but with practised calm the detritus was quickly and efficiently removed and dinner seamlessly resumed. Great imagery, and what, after all, is a technicolour chunder amongst gentlemen?

Corporal T played a starring role in my all-time favourite anecdote of my year in the mess Portakabin. I tell it with the hope that I do not appear arrogant or egotistical in so doing. He was a small, wiry, slight character, with a cheeky but dry twinkle to his eye. He had a number of gaps tooth-wise and, all in all, in his quiet, unassuming, calm way was one of those roguish individuals who one cannot help but like – Edinburgh's slightly dour take on the Sassenach capital's cheeky cockney sparrow. I certainly liked him. He was very good to me and I think was embarrassed by the conduct of some of his young officers towards his 'guests'.

One evening after dinner he approached me as I prepared to leave the dining room as the last man there. I still can't believe his words; they came so far from left side, with absolutely no preamble or build-up. He asked me, bold as brass, whether I would like to take advantage of the fact that, apparently, a number of the soldiers' wives found my boyish English good looks attractive and would like to express their appreciation personally. Corporal T stated it would be 'nae problem' bringing them unobtrusively to my Portakabin room, adjacent to the rear door to the mess, whenever I gave him the nod. An absolutely unbelievable suggestion and equally completely fallacious appreciation that such an egregious breach in military etiquette would remain unobserved. Frankly, I was knocked way off kilter but managed to politely thank him for his kind initiative and concern and told him that, on the basis that I thought his thoughtful and flattering offer was not a very good idea, I must reluctantly decline. He shrugged, and that was the end of that. It was as if he had offhandedly asked me whether I would like another cup of coffee or a copy of today's *Telegraph*.

In idle moments, I have entertained notions as to what would have happened if this plan had come to pass and, of course, been compromised. The nuclear detonation in disciplinary terms would have resulted in the court martial of the year, for sure, and my immediate requirement to find alternative employment. To parrot the phrase 'He was drummed out of the mess' takes absolutely no account of the scandal and outrage that would have followed. Luckily I possess some good sense and a well-tuned moral yardstick, even then as a tender 23-year-old.

About the same time, the very attractive and sensual wife of a fellow officer made it abundantly clear that if there was to be a queue outside my door, she would like to be at the head of it. Again, I deferred to common decency, the code guiding conduct towards one's brother officers and a healthy respect for the martial justice system. There were a number of other 'goings-on' going on within the confines of the wire that would have fed a string of juicy courts martial, but my attachment continued in a thoroughly detached, if unadventurous, manner.

I left Ballykinler ready for a change. I had learned an awful lot, had had a lot of fun, as well as aggravation, along the way, and the mission of showing me how the 'real' Army lived and breathed had been achieved, warts and all. I was now ready professionally for a new challenge. This time the challenge of a full-time dedicated intelligence job. I did not realise it then but I was effectively two steps, one very short, the second much longer and more protracted, from my posting to Berlin and Mission life.

8

On Pass

It was not long after I'd settled in to life on the corridor that I came on pass and began my first month of touring.

Teams were not fixed and would likely alter for each tour. I ended up touring with most of the Army tour NCOs and a couple of the tour officers by the end of 1990. I have to say, there was no one that I found hard to stomach for three days in the same vehicle. I am a tolerant person but I don't tolerate fools readily. I think that certainly says something positive about the unit.

BRIXMIS tours were invariably mounted with a three-man crew: the tour officer in the single back seat, the tour NCO as front-seat passenger, and the driver. All would have attended the course at Ashford and be on pass. Tour NCOs could be deployed as tour officers, if required, generating enhanced flexibility and, in my time, one – Jim E, a member of the Parachute Regiment – was even a qualified Russian interpreter. Stereotypes?

A slightly tongue-in-cheek entry in the 1968 USMLM Annual Report described tour officers:

> The tour officer is a 'one-man band.' He is a … reporter, con-artist, resource manager, … teacher, perennial student, and military representative of US interests in Communist East Germany. The tour officer must address these various roles creatively. He must be at once innovative and aggressive. His stamina, coolness, and patience must be high. Last of all, a tour officer must be lucky. With a little bit of luck, your 'average good man' will be in the right place at the right time …

> There are few assignments in the military service where an officer can personally contribute so much to US defense efforts and few jobs where a man must be such a jack-of-all-trades and master of each.

I would extend this thumbnail sketch to include tour NCOs as well, particularly those in BRIXMIS. Mission drivers, both Army and RAF, were carefully selected and were fully integrated into the unit, enjoying respect and acknowledgement as professional equals. Interestingly, the driver/tour NCOs of USMLM were drawn separately from another casting process, initially recruited from the local Berlin Brigade, then, after the Nicholson shooting, from a local Special Forces unit and recon or Ranger units based in West Germany. They were selected primarily for their ability behind the wheel rather than their capability to provide direct tour-related support to the tour officer. They received specialist training at the Mercedes driving 'academy' near Stuttgart and on the Berlin Police special operations/anti-terrorist driving course.* All their drivers were good German linguists complementing USMLM tour officers' Russian ability. US crews were unquestionably effective on tour but did not possess the breadth and versatility of their British counterparts. I am not sure how the French managed manning issues but with their smaller number of passes, they too had adopted the two-man tour crew model.

Tours followed two basic formats – what I would describe as 'standard' ones, and others denoted as 'local' tours. Less frequently mounted was a third category that I will imaginatively label as 'other'. Standard tours were generally of three days/two nights' duration, but it was not uncommon to push this out to five or even seven days when the situation on the ground demanded intensive coverage. USMLM kept theirs generally down to two days, which undoubtedly worked better with their two-man crews. In the event they mounted what they described as 'extended' tours, they overnighted in East German hotels. This level of tour luxury BRIXMIS restricted to its cultural tours, a feature of my 'other' tour designation – the prospect of requiring wives to snuggle down in one-man tunnel tents with a pack of wet wipes was obviously not a realistic

---

* Members of Joint Intelligence Staff (JIS) also attended the Berlin-based course. It was excellent. No matter how skilled or experienced a driver one is, there is always something new to learn and practise.

aspiration and defeated one of the main objectives of these higher-profile flag-waving exercises.

By the mid-1980s the main vehicle deployed by all three Allied Missions was the 3-tonne G-Wagen, a superbly competent and robust cross-country 'jeep' that probably could not be bettered in its role by any other brand or mark. The history of Mission vehicle deployment was truly an evolutionary tale almost as varied as the lineage of humankind. Humber had equipped the early days of the Mission with 'Boxes' and Snipes. By the early 1950s, Opel Kapitäns had replaced them, initially painted black until the end of the decade when khaki green was adopted. The Kapitäns became Admirals until the Senator was introduced.

Our G-Wagens were equipped with safety roll bars, just in case, and rally seats enhancing comfort and stability for the crew on choppy flights. If I could afford it, this is the SUV of choice I would drive. Land Rover Discoveries and then Range Rovers had been trialled and found wanting. They were just not tough enough – a sad testament to the UK motor industry, as was. On the course back at Templer, we had used the former. I remember sitting on the hard shoulder of the M20 enjoying the sun whilst we awaited recovery after yet another breakdown. On the other hand, the Mercs were veritable tanks and proved time and again that they could handle the toughest challenges on and off-road, in the hot summer and in the midst of a hard central European winter, in the dry, the wet and the snow. Some of the Soviet exercise area tactical routes, 'tac routes' in Mission parlance, were extremely tough and unforgiving but really little challenge for the rugged G-Wagen.

BRIXMIS had fifteen of them. They had replaced the fleet of Range Rovers that had come into service in the 1970s – a perfect about-towner for footballers' wives and positive, demonstrative support for the UK motor industry, but a failed compromise. They were mechanically unreliable; had a critically poor range; drank fuel like a diabetic wino; and were poorly insulated: freezing inside in the middle of winter and draughty and suffocatingly dusty for the crews in summer. The Range Rovers had been supplemented by 3-litre Opel Senator saloons in the early 1980s. They were equipped with retro-fitted Ferguson Formula four-wheel drive, reinforced suspension, large-capacity 180-litre fuel tanks and half a tonne of belly armour to protect the underside of the vehicle from rough and uneven terrain – in no shape or form a production model. They had nowhere near the off-road capability or the space and comfort of the G-Wagen, but were sleek, fast, projected a lower more discreet profile and could still competently get the job done in all weather and over most terrain. I toured twice in one, breaking down and having to be recovered on one of these tours, and missed the space and relative comfort of the G-Wagen.

The average life expectancy of Mission vehicles was around six to nine months, by which time they would have completed their allotted 90,000km. Frankly, given the nasty and brutish nature of their lives on and off-road, this was not an extravagant fleet management programme.

The US Mission had always used US-made vehicles, Fords and Chevrolets, until the early 1980s when they could no longer turn a blind eye to the reality that their home-grown quality was surpassed by German auto engineering. Made for the prairies of the Midwest, they simply were not robustly enough designed or manufactured for the tracks and minor ways of eastern Germany. They too then drafted Opel sedans into the fleet before acquiescing to the tri-Mission strategy of conformity and the adoption of the G-Wagen.

Each vehicle in the Mission fleet was fitted with a unique number plate embodying the Mission shield and the car's BRIXMIS number. The Americans and French used similar plates. The Chief always travelled in vehicle number one. All vehicles, regardless of mark, were fitted with a system to isolate all internal and external lighting to enhance stealth, even to the extent of replicating the profile of a motorcycle with a single headlight. At the lower end of the tech scale, an array of curtains limited observation into the vehicles when the team were busy on task, and critically, sensitive and accurate trip meters were fitted, which proved invaluable aids to navigation, especially off-road through the dense forests and on tac routes. G-Wagens were equipped with infrared spot lights and another extremely practical after-sales add-on, an electric winch bolted onto the front bumper, a key ally against the cleverly disguised mud pool that would trap even an unwary Mercedes SUV.*

The tour plan, the mission of and tasking for each tour, regardless of whether it was air or ground, was drawn up by the ops staff, the SO2 G3

* As part of one's leaving gift on posting from the Mission, the recipient would receive a number plate and a metal model of a Mission car on a small plinth. As military memorabilia goes, they were special and highly distinctive mementos. When the final end came, everyone then serving received these farewell gifts.

supported by his SO3 – in my time, the assistant ops offr, as I have already noted, was also Int Corps and became a good friend. We ended up sharing one of the most unique operational experiences and claims to military fame, but that was some little time off. The tour plan outlined the fixed targets, including TOMAHAWK sites, the tour would 'visit', and added any additional tasking.

Once they had a copy of the tour plan in their hands, the tour officer and tour NCO would, as part of the doctrine of command leadership and the conventional military cascade, drill down and plan the details of the next few days, not only the transit routes moving from location to location, but more poignantly researching each target from the standpoint of likely 'enemy' activity there, equipment based there and how best to 'attack' the target from the perspective of routes in – likely points from which to observe activity based on factors of stealth and effectiveness – and routes away. As I alluded to earlier, each target had a corresponding target pack that presented and recorded the Mission's history in relation to that installation, training area, dump or airfield. The more important targets had corresponding packs that were decidedly meaty, populated with a tremendous amount of critical detail. Updated by the tour officer after each visit, packs included relevant maps, photos and sketches to provide teams with the most up-to-date physical picture of the target and key points of geography and topography relating to it. All visits to the target were recorded by date and provided a summary of activity and key observations and any relevant incident or event that potentially affected the security of an upcoming visit. On this theme, the packs elaborated on activity relating to sentries, their state of alertness, their reaction to tour vehicles and any other factor that tours had to factor into their planning and then their operational approach on the ground. Best routes approaching the target were detailed – these were also obviously relevant as escape routes too if tours needed to vacate the immediate area rapidly.

I had heard there were times in earlier days when this collaborative approach to tour planning was certainly not the order of the day. Some individual tourers religiously kept their approaches to targets and the locations of OPs very much to themselves so that the risk of fellow tour officers inadvertently compromising them to the Soviets or the *Stasi* – were reduced. This type of dog-eat-dog, *prima donna* environment must not have made it a pleasant corridor to work in but, like the Humber Snipes, this was no longer a factor of Mission life.

To supplement these comprehensive data mines, only a fool would not take advantage of picking the brains of his colleagues to assimilate their experience and assessment of a target. And additionally, the Mission maintained a real intelligence collection treasure: a marked wall map covering the whole DDR, visually displaying the very ground truth of operational reconnaissance in-country. Constructed from 1:50000 sheets, it stretched not far short of 100m snaking along the curvature of our corridor. It graphically displayed all our target locations, these very observation points and a plethora of other tour-related detail including all barracks locations, locations of dumps for TOMAHAWK-ing and emergency deployment areas. It was an incredible living, breathing tool and an impressive testament to the professionalism of the unit. I am sure the map was saved and is protected somewhere but I have no knowledge of that location personally – perhaps most likely archived in the Imperial War Museum in London. I hope so.

Routing out on tour was very much an exercise in time and distance estimation which, in itself, was a factor of experience. Normal planning criteria did not apply – distance had to be assessed in terms of the road and ground conditions that the tour would meet and was tempered to some extent by the assessment of the contingencies the tour might encounter as it made its way en route. Thus, routes were measured in time taken, not distance travelled.

On the face of it, this quick description may not vary too much from the sort of approach to planning a touring holiday, moving leisurely from the coast, perhaps to a remote church, then further into the mountains before hitting the mediaeval provincial capital. Unequivocally wrong, however. It was a little more complex and focussed, and the implications for inadequately considering all the factors were a little more serious than getting back to the seaside to find one's favourite restaurant had just closed.

The reality was not only long distances to cover – often over tough terrain and questionable roads, not inconceivably clocking up 2,000km for a comprehensive three-day stint – but at the end of a long day, the requirement to execute additional critical, key tasking. Evenings were spent not erecting tents, building camp fires, then sitting around them chatting or singing. Evenings were spent tucked away surreptitiously observing key rail lines or, in the unlikely event of their absence, nodal road junctions, in anticipation of movements of massed tactical equipment either in transit from or to training grounds, or of course that fateful bowel-loosening preparatory move before crossing the IGB. The reality was that precedent had confirmed that most rail moves took place at night, no doubt to increase security but also in likely recognition of the priority of civil traffic that would monopolise daytime rail movement on the East German national rail network. Tours would sit shrouded in darkness focussed on the line waiting to get cameras clicking to record the flatbed passengers, image-intensifying Moduluxes fitted to lenses, transforming a 35mm SLR camera size-wise into a mini rocket launcher, once photographic light levels dropped sub-critically. At some point in the evening the team would then cook dinner and eat.

This task, always keenly anticipated, would also be undertaken on board within the confines of the vehicle. A small burner would heat a large cooking pot precariously balanced between the driver and the tour NCO into which each of the three members of the crew would pour in either parboiled rice, vegetables or meat, and sauce to construct, invariably, an *ersatz* curry or stew – the art of one-pot cookery epitomised. The menus would be decided upon in advance and the makings brought from home; who would bring what was decided upon as part of the tour planning – Campbell's meatballs were a mainstay and could always be found on the shelves of the NAAFI. Very soon the pot would be simmering away, filling the vehicle with a more acceptable aroma than the odour of stale bodies and that indefinable 'military smell' that seems to permeate all things martial. In midwinter the stew/curry would be ingested and stimulate a degree of internal warming, but very soon the ambient temperature would drop again and make the donning of another layer of clothing imperative. It is simple rituals and the satisfying of basic needs that, at least for me, have always been the catalyst for a realigning of expectations, objectives and desires that made soldiering, and touring as an aspect of the greater part, fulfilling. In the heat of summer the reverse process would occur, and tours would most likely be poised in short-sleeve order, to use the vocabulary of our military dress code. But at the very least, the ritual always served as a welcome distraction and provided something tangible and comforting to anticipate in order to punctuate the long days as they wound on towards midnight, but by no stretch of the imagination to bedtime.

As in any task undertaken by the Missions, once a third party had registered the presence of a tour vehicle, it was invariably time to relocate one's position. At this stage of the Cold War, it had to be assumed that the vast majority of potential observers, if not military, would certainly be likely to make a report to the authorities and provoke a potential detention; however, for me the true level of support freely or grudgingly given to the state was always a matter of speculation and debate. The unquantifiable reality was that a great many DDR citizens were heartily disillusioned with Soviet occupation and the shortcomings of their own brand of socialism, but they were not, of course, readily identifiable through a reinforced Perspex vehicle window.

All tours hoped and prayed that dinner, this critical event in the daily programme, would not be interrupted by a military movement or, for that matter, by any other movement. Should this occur, on the battle cry and rallying call of 'Kit', the pot and burner were rapidly dumped, but with care and speed, out the door to be retrieved later, and the team would seamlessly move into the overdrive of the intelligence collection machine that they were. Whilst the driver kept all-round observation to ensure security, the tour NCO, sitting in the front passenger seat, would 'call' the details of all sightings into his small Dictaphone – equipment type, variant, distinguishing mark, VRN if applicable and any other pertinent detail. All equipment had both a vehicle registration number, and in the case of armoured pieces like tanks, APCs or self-propelled guns, a side number; it was this data, again as I previously noted, that the Research office regarded as its daily bread, recording it in the database and using it to subordinate equipment and identify formations.

In the case of a rapidly moving road convoy or a train passing in front of the darkened G-Wagen, waiting patiently in 'ambush', there would be a high-intensity and stressful burst of activity rocketing the crew from zero to maximum output in seconds. Simultaneously, the tour officer, from his deliberately recessed position sitting in the single seat firmly anchored in the middle of the vehicle, would be furiously photographing targets, mirroring the calls of the tour NCO so that each item of kit was referenced by a verbal description and photographic evidence – thorough, systematic and effective. In these days of wet film, after thirty-eight exposures the film had to be changed and the camera reloaded, adding to the workload and the collective stress level. What a luxury digital photography generates with an almost inexhaustible limit to finger clicking. The driver would likely aid the tour officer, reloading cameras for him as he continued to watch front, back, left and right for possible compromise. Each tour ventured forth with three camera bodies (the Americans preferred routinely to carry five, presumably so that the two-man crew had the potential to take just short of 200 exposures without the rhythm-breaking routine of having to change a film), a span of lenses from 85mm up to a 1,000mm fixed-length mirror lens, plus doubler, and a photo shop's supply of film of differing speeds, with faster films for tackling targets in situations of lower light, which we would routinely 'push' further, squeezing them to the limit up to 6400 ASA. A large convoy or train move could include several hundred assorted vehicles, so it was a frenetic event in the day's routine. And it was not uncommon for a tour, at least on a rail watch, to have more than one rail move to 'attack'.

With relation to kit trains, once a sighting had occurred, the tour was mandated to remain 'eyes down' for another 3 hours to cover the contingency of this second move, then another 3 hours if a second move occurred, or a third or fourth. This SOP was based on the appreciation that it took the Sovs, on average, a period of 3 hours to load a column of kit onto a train, driving the vehicle up onto a loading ramp sitting level with the top of the rail flat. Transloading wheeled vehicles would be relatively straightforward but manoeuvring tracks would require skill, steady nerves and certainly the absence of bad luck to get things done. The process required patience, which required time, hence this strict time-based convention.

The clock would now be advancing seriously towards the following morning, but there would generally be one more operational task to undertake before sleep. And this task was the least savoury and most unsanitary that any intelligence collection team in any theatre on any mission would have to perform, I suspect then, at the height of the Cold War, or since: OPERATION TOMAHAWK.

Soviet forces followed robust security procedures scrupulously, I suspect because the sanctions imposed on those who committed breaches were severe. If something was formally classified, it was properly accounted for and properly disposed of when it was no longer required. However, outside of the formal rules governing security, they were no different from any other organisation composed of human beings. People are essentially lazy – they cut corners, they are happy to get away with what they can get away with – and I suspect fly-tipping was an aspect of life back home where our concept of prompt, systematic rubbish collection by a local authority was as alien in the USSR as a satisfactory regular annual wheat harvest. It was this appreciation of human frailty that empowered TOMAHAWK, or SANDDUNE as the Americans dubbed their parallel operation.

Both operations were the catalyst for hundreds and thousands of intelligence reports over the years, complementing the more orthodox and 'conventional' reporting that the Missions disseminated, informing the wider intelligence community. Whilst most reported on small pieces of the jigsaw, TOMAHAWK did fill some strategically important gaps in the NATO collection plan. In the early years of the Soviet entanglement in Afghanistan, the Mission was able to confirm its use of chemical weapons against the *mujahidin*. During another of these officially sanctioned 'treasure hunts', documentation, later corroborated through more orthodox collection means, had been picked up that, when fully analysed by the team back in the Mission building, revealed startlingly that the Russians' new composite tank armour provided a degree of protection in excess of the value estimated by Allied technical intelligence analysts, thereby revealing our current shoulder-fired anti-tank weapons were completely ineffective against it. Intelligence of this quality at this level, whilst rare, was fundamental in guiding research and development programmes and potentially shaping and steering, confirming or precipitating amendment to tactical doctrine.

OP TOMAHAWK was a carefully, and scrupulously well-protected intelligence source. As a collection tool, on the scale of technical complexity and sophistication, research and development, and investment of resources, it was about as far to the left as one could go: cheap, incredibly simple and low maintenance. TOMAHAWK was a clandestine scavenging effort that systematically combed rubbish dumps adjacent to military installations. So carefully controlled and executed was the operation that at no time during its conduct was it ever compromised, within the Allied camp, and the source of the intelligence identified. However, those same Soviet intelligence agents who revealed the extent of Mission collection are very likely to have compromised 'rubbishing' as the invaluable source that it was. Even though this may have been the case, it did not appear to have constrained GSFG/WGF's SOPs governing rubbish disposal.

The physical protection of this source – no reference was ever made to it on the course at Ashford – was maximised operationally by conducting rubbish collection in the quietest hours of the night, generally between 1 a.m. and 4 a.m. before first light even on the longest days of the year. Discussion of the operation was controlled even within the Mission building, and especially in the House, and those engaged operationally conducting collection were limited. Not all tourers were tasked on TOMAHAWK, just the majority of part-timers.

So with the chimes of midnight a distant echo, tails clear of any lingering MfS surveillance, whilst the driver remained with the tour vehicle, hidden some way off from the target, the tour officer and NCO would move with utmost caution and discretion onto the dump. Details of dumps were recorded in target packs in the same way all other targets were logged, and each had their place on the master wall map. Dumps could be very small and localised or cover acres. At the dead of night, without using light, wearing standard-issue flying boots covered with rubber over-boots to avoid leaving military tracks, the team would move systematically over the remains, collecting anything that appeared to be of value. This could be documentary, such as letters from home confirming unit designations and addresses, pages from signals pads, training manuals, notebooks, technical manuals or discarded bits of hardware. Quite conceivably the pages in question may have been used in lieu of toilet paper as this was one luxury too far for the Red Army conscript and was not issued for field use, unlike the small pack tucked away in the field rations of the British soldier. He therefore had to reach for anything made of paper that happened to be close at hand at this critical moment of the day before it ended up on the dump – no pun intended.

To bolster these routine pickings, it might also include all the items that the regular turnover of conscripts, at the end of their time with GSFG/WGF, threw up. These accumulated possessions would be dumped rather than forwarded back to the USSR. They were definitely productive times. However, if more routinely, a recent, fresh delivery of rubbish could be identified so much the better. The chances of finding something exploitable rose exponentially. Likely items, once identified, isolated and removed from their unpleasant and unsanitary environment, were placed in an innocuous East German gunny sack.

The tourers carried nothing on their persons that, if dropped, could indicate any Allied presence and removed the Mission badges and rank slides from their olive-green non-standard uniforms.*

You can imagine, midsummer, the environment. The smell was pretty awful: strong, penetrating, lingering, seemingly impossible to quench from one's hands and clothing. Footing was treacherous in all seasons. There was the continual pulse of maggots, the buzz and hum of other insects, too dark to identify, and the scurry of rats and other animal scavengers, sometimes including packs of wild dogs. I did not encounter this particular worrisome hazard. And there was that bizarre warmth that seems to emanate supernaturally from deep within piles of living garbage. The chance of a wound from anything, pointed, sharp or jagged was ever-present as we poked and sifted. The thin rubber gloves we wore offered zero protection, but anything more robust would have seriously impeded our ability to sort and

* Mission uniform was strictly paramilitary and a matter largely of personal taste, within certain generic style guidelines. Stating clearly for the record, Mission tours were armed with nothing more lethal than forks and spoons, excepting of course the ubiquitous multi-tool. I wore a standard green Army long-sleeved shirt. If it was midwinter, I would wear a round-neck, long-sleeved RAF flying top, or thermal underwear as a base layer, and I might substitute a Norwegian zipped-collar shirt for the long-sleeved button one. Over the shirt, again in winter, I wore a standard Army woolly-pully jumper, then a US M65 Vietnam-era olive-green combat jacket that I had picked up years before on the Edgeware Road. Trousers were the standard issue green 'lightweight' trousers with thigh cargo pockets, again maybe on top of thermal long johns. It was not unknown for British soldiers, in the height of winter, to wear ladies' tights – light, comfortable, cheap and warm, apparently. I have even heard of Royal Marines wearing stockings and suspenders but there were no Commandos in the Mission in my time. Boots, as I mentioned, were all leather RAF flying boots, with their non-military tread pattern. Again, at the height of winter, some of the guys wore locally purchased East German 'engineer'-style leather pull-on boots with a warm lining – I never bothered. For really cold weather the Mission issued Canadian parkas complete with fur-lined hood. Gloves were another personal choice. Tourers frequently used green leather flying gloves inside the vehicle to help reduce the 'glare' that white hands clasping cameras or other equipment radiated. We also had an issued pair of Arctic mittens. I was not a big fan as, personally, I dislike the sense of restriction that thicker gloves give, but outside in the extremes of winter 'proper' protection was frankly essential. For winching operations the crew were issued heavy-duty rough leather 'work' gloves to protect palms and knuckles. We wore plain rank slides and Velcro-affixed BRIXMIS badges on both upper arms, where Boy Scouts wear their merit badges, with their distinctive Union flag design. Again, for that biting winter weather, we had shapka-type ensembles with pull-down ear flaps but I think I tried mine once, looked in the mirror and never touched it again. We took berets with us just to wear on crossing the bridge as a more formal acknowledgement of our Russian hosts. In line with my nonsense 'cover', I wore the dark-blue beret of the RCT with its embroidered officer badge positioned over the left eye. Our Soviet opposite numbers were equally dressed down and informal, but the Americans, after the introduction of their DPM BDU, took the decision to wear them. Psychologically, set against the Soviet military mindset, disruptive pattern camouflage gear equated to special forces.

select, especially pieces of paper. These were all matters for serious concern. Some dumps would include medical detritus – used and soiled bandages, phials of medicine and drugs and, of course, the potential for used hypodermic syringes. I heard stories of one particular tip, adjacent to a military hospital, frequently hosting amputated limbs as casualties from the war in Afghanistan were brought to the DDR rather than directly home to the *Rodina* in order to minimise ammunition for the anti-war movement there.

The winter scene was better in some ways but items were frequently frozen together, making it hard to assess their intelligence value, and of course isolate them for collection. Then movement was even more treacherous, and it could be debilitatingly cold, with temperatures consistently dropping to -15 or even -20°C below zero in the dead of night. At least the cold was generally dry and not that damp variety that penetrates to the bone.

Paradoxically, however, I rather enjoyed the TOMAHAWKS I conducted. The task was certainly absorbing – time passed very quickly, even though a thorough rummage might last two or three hours at least, and I was always excited to return to Berlin, to the Spandau office, to find out what we had 'liberated' – as I have already said, this was Spandau's sole *raison d'être* by 1989. They processed, translated, collated and onwardly disseminated the tours' finds. I don't remember returning with any notable scoop, but I'm sure not all my efforts were completely wasted. I do, however, as an aside, recall 'rubbishing', but not on an established tip, a couple of pages of a communications log left behind after a training exercise had ended. Apparently it contained some pertinent intelligence for the SIGINT community.

I experienced only one drama on my various tip trips. On a smaller site close in to the rear of a barracks, it was possible we had been heard making our way from its edge in towards the opposite corner that we had pre-targeted for exploitation, but I suspect it had, more likely, been something else less stealthy than the two of us that had alerted the installation sentries. Regardless, as we pulled and pushed, sifted and sorted, we in turn could hear the noise of them approaching unsteadily but consistently. We promptly decided to beat a hasty retreat, grabbing our sparsely filled sacks and making haste back to the tour vehicle as rapidly as circumstances allowed. Being detained on a tip *in flagrante* would have been a devastating tactical error of far-reaching proportions, so we had absolutely no qualms about running away. After all, he who runs away does live to rubbish another day. There was always the chance to return on the next tour – our dumps were the only 'living', lasting evidence of Soviet occupation that remained behind in the old DDR after reunification and their withdrawal home.

At the end of a night's operation we scavengers would cautiously retrace our path back to the waiting driver sitting nervously alone in the vehicle at the RV point. As an integral contingency briefed in before commencement of the task, an alternative RV was identified if the driver should, for any reason, have to move. But once back at the G-Wagen, the evening's takings were slipped, sack and all, into bigger plastic bags to contain both stench and bacterial risk, and the crew would exit the area with enhanced tactical care to make for the pre-designated 'Z-platz'. We ensured there were no trailing lights from other late-night revellers, happy that we could now grab a few hours' much-needed sleep in our one-man tunnel tents after using our personal supply of wet wipes to de-gunge. This was the downside of the deal – one had to accept that, in reality, we were off to sleep as unsanitary as it was for a human being to be, short of touring a sewer and wallowing in the contents. That unforgettable cocktail of foul odours is one that does linger for some time, both physically and in one's imagination.

Z-platzs were always located well inside larger woods, where the likelihood of casual detection was remote, especially as we were always early risers. The driver slept in the vehicle, adjusting seats to achieve the most comfortable position. The tour NCO and officer slept outdoors, *al fresco*. The tunnel tents were partially assembled and collapsed in the back of the G-Wagen so just required the bare minimum of adjustment, once positioned on a suitably flat rectangle of forest carpet, to be usable. With an insulating mat and Arctic down sleeping bag, sleeping fully clothed, we could usually ensure a relatively comfortable few hours of sleep even in the frozen depths of December and January. The extreme fatigue of a day's hard touring was a forgiving bedfellow.

I have two lingering memories of nights in the great DDR outdoors: the size, ferocity and extreme itchiness of the mosquitoes and their bites during summer months and my panic in the early hours one night as I awoke with the fabric of the tent pressing down on my face, smothering me. I had failed to secure the four corners tautly enough and it had collapsed, undoubtedly as I turned in my sleep at some ungodly pre-dawn hour. It was one of those horrible claustrophobic moments when one is too fatigued to understand what is happening but conscious enough to fully experience the irrational sense of fear.

There was one instance during my time when a tour team likely brought back something completely unexpected and, arguably, potentially life-threatening. The tour officer and NCO in question, both of whom I had toured with, were both struck down with what progressed into a long drawn-out and sapping medical saga. They fell sick together, and both, interestingly, exhibited identical symptoms. This included long-lasting extreme fatigue. These were the days when there was a heightened awareness of ME, but the fact that two of the Mission's personnel were affected simultaneously seemed to stretch the bounds of coincidence. To cut a long story short, I subsequently heard after the Mission had been disestablished and the pair posted on that, after extensive testing and diagnostic pontification, their malaise was tied to a TOMAHAWK the pair had completed together. It was concluded, so the telling went, that they had been subjected to discarded chemical ordinance, or a constituent chemical component thereof, either linked to tactical use in Afghanistan or, alternatively, and to my mind more likely, sub-lethal doses used in exercise simulation in-country within the DDR. If true, pretty scary stuff, but entirely feasible given the time, the realities of real Cold War competition and what we did under the auspices of the operation. I am not sure if they made a complete and full recovery, or whether there is any negative postscript. I sincerely hope not, especially set against the precedent of 'Gulf War syndrome' and the absence of support that sufferers failed to enjoy from HMG.

9

Touring

Each and every BRIXMIS tour shared a number of identical features regardless of targets, duration or type. They went out and returned to base in the Stadium complex via the US Mission located to the south of the city, firmly in the American sector in the upmarket district of Zehlendorf. Interestingly, from the standpoint of historical continuity, USMLM had been housed since 1957 in the Nazi wartime Albert Speer-designed German Military Supreme Command headquarters run by General Keitel. Prior to the US Mission taking occupancy, the Office of Strategic Services had taken on the tenancy following Keitel's forced eviction and had passed on the deeds to the Berlin CIA station. Very conveniently, it was just across the road from the PX shopping centre. Apparently the building now houses a *kindergarten*.

At tour's end, on the way home to the Olympic Stadium, the team swung by to record a brief highlight of significant sightings or any other noteworthy event experienced during the tour and enjoyed access to a fridge in the debrief room stocked full of typical snacks from the land of the free. After a few days of boil-in-the-pot, generally healthy but tedious mediocrity, a quick inject of high-calorie carcinogens was welcome. The French may not have played, I suspect, because of limitations of language, but I honestly don't remember and certainly never bumped into a French tour during my numerous visits to USMLM.

Outbound then, all tours from all three Missions conducted themselves over and across the Glienicke Bridge, the infamous exchange point for Cold War spy swaps. For us, this was the single point, and had been since the building of the Wall, where the Soviets could locate tours: they knew where a tour absolutely had to be at tour's start and would categorically return to at tour's end. The Bridge literally marked the frontier between the city of Berlin and the provincial city of Potsdam, the capital of the eastern state of Brandenburg and the border between east and west. It was a robust twin-span construction and quite possibly one of the least-used bridges in the world. The only other Bridge crossers were diplomats who had business inside the DDR outside of Berlin. Tours would pause at the barrier and display their Mission passes to the Red Army officer of the watch. The tour crew's identities would be recorded and they were then free to slip the forty years of technological advancement that marked the superiority of western civilisation and enter the sombre sameness of DDR socialism.

The contrast that the passage across a short bridge over a narrow piece of water could represent was both graphic and disorientating, but somehow perversely calming. Despite the crystal-clear signal that a period of hard, tiring work was about to commence, the passage and return to a simpler technological age and, for BRIXMIS, one minus the intrusion of any form of two-way communication was soothing. One really did enter another land, as part of another world, at another point in history – in a word, a parallel civilisation different in every conceivable way. It was that stark. The only other time I have experienced so surreal a sense of time travel was walking across the border into southern Albania from Greece in 1999 at the start of the Kosovo War.

It is still not infrequent to catch images in various TV documentaries of shots taken over the Wall from the west into the eastern sector of Berlin. They display that energy-sapping, soulless greyness that so typically characterised not only East Berlin but all other large conurbations in the DDR. It took one a moment to spot that the impression of grey was magnified by the absence of any other significant or vibrant colour. The DDR was essentially pigment free and the reason, with a moment's reflection, was obvious – paint was a luxury, an unnecessary extravagance that diverted valuable scarce oil resources from the heavy industry and martial applications where they were most needed. There were few bright lights, little evidence of the aggressive in-your-face advertising of the West, and few enticing shop fronts. It was a land of opposites. Somehow the melancholy of houses, blocks of flats and all other buildings, and the very streets themselves, was supported by the drabness and spartan utilitarianism of the cars, lorries and buses, and the pallor and styleless clothing of the people. But despite this impression I am painting of a depressing, sad and listless reality, not everything was negative. There were spectacular towns and cities, rich in architectural and historical splendour, where the absence of aggressive consumerism and ugly modernity enhanced a more simple purity and beauty. I am remembering places like Dresden, Leipzig, Halle, Erfurt, Eisenach, Weimar, Bautzen, Görlitz and, of course, Potsdam itself. The DDR was dotted with beautifully unspoilt villages projecting the very essence of historical continuity. Many had survived the ravages of the Second World War undamaged as the Soviet Army had bypassed them in the race for Berlin and then onwards to points west in the competition to occupy German territory. Nature abounded. Particularly evident were the raptors majestically soaring above the fields, here in the east paradoxically hunting prey protected by the absence of the randomness of the harsh toxic pesticides used in the west. Wildlife co-existed in harmony with man, where the landscape had not been despoiled by the ravages of industrial farming. And not all the people regarded us with hostility or suspicion. When we stopped to augment our diet with ice cream or cakes, and to make other small purchases, we were not infrequently met with smiles and enquiring and engaging openness.

Clear of the brief formalities at Glienicke, berets now returned to bags, tours then transited across town to their respective Mission Houses to check in before launching out into the greater East Germany.

The BRIXMIS House, 34–37 Seestrasse, was a very short drive just around the corner, off Berlinerstrasse, the main Route 1, backing onto the Heiligersee, a relatively small but pretty lake – the French House was a very near neighbour, literally a handful of residences further along the street. The Mission House was an imposing and majestic whitewashed villa, I think the most striking and imposing of the three Mission Houses. It was flat roofed, three storeys high with a stunted, clipped wing on each flank. The front entrance was covered by a four-pillared portico with a balcony accessed from the large windows on the first floor. It would not have looked out of place as the official residence for a first world ambassador in an important partner country. Of course the Union flag fluttered imperiously from a flagpole mounted on the roof. A further flag staked claim to the garden and displayed ownership for all to see on or around the lake. This was our official HQ, a sort of quasi-consular facility. BRIXMIS had been 'living' here since 1958 when the original premises had been severely damaged during the riots that swept through East Berlin and beyond. The house had been built around 1890 and owned by a Jewish family who had fled to the US in the 1930s. It had remained in state hands as a teacher training language college until we took occupation. Apparently the teachers had been evicted unceremoniously and given a mere three weeks' notice to relocate by their Soviet landlords. It is now incidentally home to Wolfgang Joop, the renowned parfumier and fashion designer, a native Potsdamer.

The front entrance was 'protected' by a sentry box, the occupant – without doubt a *Stasi* operative – reporting all movement by telephone to the guard room on the bridge. The manicured gardens to the rear of the house sloped down to the lake with a clear view across the placid water to the National Army Museum, and the old holiday cottage of the von Stauffenberg family, of anti-Hitler 'July Plot' fame. The property was all quintessentially British and projected something strong and enduring in the imagery that played such an such important role in the propaganda battle that raged throughout the duration of the Cold War. In marked contrast, at the north-eastern corner of the lake, local Potsdamers enjoyed their only true expression of freedom, a *Freikörperkultur*, FKK, nudist beach – a shock to me when I stumbled upon it for the first time when out for an evening stroll from the house.

Also within easy walking distance was the beautiful magpie mock Tudor Cecilienhof palace where the Potsdam Conference had been held in August 1945. We could go there for dinner, in uniform of course, and eat and drink in style for a pocketful of small change. The cheapest beverage on the menu was Soviet champagne from the Crimea. It was always an impressive and somehow sublimely relaxing, calming experience contrasting with the high-octane intensity of socialising back in the west. One could not help but feel and almost smell and taste the weight of history hanging from the chandeliers and clinging to the wood-panelled walls.

The Mission House was a fantastic venue in which to host visiting dignitaries and military guests, Allied and Soviet, particularly in the summer months when the garden came into its own. The lawns were the perfect venue for garden parties, for strawberries and Pimm's, or strolls to take in some air during dinners and balls staged in the House. If there wasn't a croquet lawn, there should have been. The one note of caution stressed to all visitors and Mission members alike was that the staff were in the employ of the *Stasi* and the KGB. All calls made on the external phone line were routinely intercepted, and there were also quite possibly a scattering of eavesdropping devices throughout, at least in the public rooms of the House. Across the lake, there was an observation post periodically manned in the clock tower of the museum, possibly armed with directional microphones. As I was to uncover once the Wall had fallen, in the event, suspicions concerning listening devices were unfounded but the staff were indeed regularly debriefed, and telephone calls systematically intercepted, logged and collated as part of a full-time operation run by the *Stasi*. The need for caution and discretion was imperative. Despite these dystopian admonitions, the atmosphere in the House was always light and cheerful, and I experienced no problem or stress in carefully picking my words or conversation points.

I particularly recall the formal ball we hosted on 21 July 1990, after the Wall had toppled. The date is not hard to remember because it coincided with The Wall concert when Pink Floyd came to town to make their mark and add some melody and rhythm to the lingering celebrations. In contrast to the scruffy thousands who flocked to the site between Potsdamer Platz and the Brandenburg Gate, our gathering was by invitation only and honoured our American, French and Russian brothers in arms, plus the who's who in the British zone. It was a brilliant occasion, medals flashing, the grandeur of a bygone age reflected in the cut, style and hues of our mess kits and the ball gowns and sparkle of wives and girlfriends.*

Those non-members of the Mission fraternity had been bused across the Bridge, complete with Mission 'tour guide', with the blessing and authorisation of SERB, which was our interface with the Soviet Command. On an occasion previously, I had taken on the role and it was actually good fun. No doubt they remember this July night as keenly and clearly as I do. The evening really had taken on all the glamour and substance of a court ball in imperial days, long gone in St Petersburg, Vienna or even Victorian London.

Our Potsdam Wall event came hard on the heels of an annual event at the other end of the sophistication scale, no pejorative allusions intended: the

* I could repeat the tired old anti-Royal Air Force dig that irked some members of the RAF in these days before the post 9/11 medal bonanza. Against the GSMs (General Service Medals), the ACSMs (Accumulated Campaign Service Medals) and other campaign medals and awards worn by Army tourers, the vast majority of junior service members were bare breasted – NFM, No Effing Medals.

Fourth of July celebration at USMLM's house. This was a great fun day of steaks, burgers, horseshoes, weak beer (again no pejorative allusions) and various boating contests on the adjacent lake – it was homely, down to earth and honest as only the Americans can serve up. The highlight of the day seemed to be a long-running fixture, pre-dating its contemporary popularity, a sort of beach volleyball competition. The key game was always the match between the visiting SERB side and the hosts. Neither side readily accepted defeat – superpower Olympic rivalry writ small. Looking back, we were a very tight community united in a single, unique purpose and mission, equipped with resources fit for purpose, ready to take on tasks and opportunities mirrored nowhere else in the military orbit. I think at the time I, and undoubtedly others, took these things for granted and failed to 'live the moment' to the full on too many occasions. Way, way too late now but I do regret it, and pushing aside my natural cynicism and professional arrogance, I do miss occasions like these Independence Days. Using the word in its purest sense, they were unique.

The last remaining point of touring commonality was the limitation that PRAs imposed on tour movement. Encroaching into a PRA and being apprehended, or at least photographed doing so, would lead to serious implications. Official complaints would be raised and presented by SERB, requiring official responses. This is not to say there were never deliberate transgressions – PRAs did after all provide protection for key installations and training areas – but these were scrupulously managed and executed. More *ad hoc* was corner cutting to avoid lengthy detours – tours would move smartly across a 'corner' of a PRA for perhaps a few handfuls of hundreds of metres routinely – I did several times without any issue or indeed much concern.*

A more interesting issue was the topic of TRAs, slapped down with little warning, relatively infrequently but with sufficient regularity to disrupt touring continuity, to provide cover for a short-term reason, usually training related. The British Mission dealt with these ploys to frustrate our intelligence collection in an ingenious way. I presume the other Missions employed a similar sleight of hand given the close liaison, but I have to say I am not certain of this.

---

* With the demise of the Missions, when the role passed to JIS, we certainly paid absolutely no regard to any restriction maintained vainly by the Soviets or any other authority. I will outline our role and operational remit and what we achieved later.

The first point to make is that all tours, once they entered the DDR and moved on from their brief stop at the House, did so without any form of two-way communication with the outside world, or specifically back to West Berlin, as I previously highlighted. Teams carried commercial radios to tune into *The Archers* and *Woman's Hour* but had no means of initiating any communication. Had military communications equipment been deployed, it would have had to have been long wave and necessarily bulky with a distinctive large whip antenna. The late 1980s, of course, were pre-mobile phone days without the added luxury of compact and effective sat com systems that were at the heart of communication during my days as a Military Intelligence Liaison Officer (MILO) ten years hence.

The reality of the isolation that this situation engendered was, for me, highly attractive. Tours were free from the direction and interference of the chain of command. Once we'd left West Berlin we were free agents and masters of our own time and tasks, if not our destiny. We made our own luck independent of others; we made our own decisions in isolation and, consequently, success or failure was ours and ours alone. In reality, however, if a tour experienced an emergency, it could call the Mission House on the East German telephone network, or even *in extremis* request help from the local Soviet Kommandatura. We were, of course, in reality, accredited to their HQ and command structure located in Zossen Wünsdorf to the south of Berlin, so we were to some extent, perversely and massively stretching the point, on the same side.

This point made, however, there was actually a means by which a simple message could be communicated to a tour. In the event a TRA was suddenly imposed by GSFG/WGF command, the Ops office in the Stadium would initiate a call to a pager carried discreetly in each BRIXMIS tour vehicle. This clandestine signal would immediately precipitate the team suspending their pre-briefed plan and pausing to open the case of the bleeper. Inside would be a simple concise printed message. The message consisted solely of a map grid reference. To this location the tour would now head, without delay.

Simultaneously the standby tour crew, back in West Berlin's Olympic Stadium complex, would be scrambling to deploy, receiving a detailed briefing from Ops regarding the imposition of the TRA and its likely implication in terms of activity on the ground and, by implication, collection opportunities for the tour across the bridge.

The standby tour would then proceed to Glienicke, mirroring any other routine tour, comply with formalities in the normal way, hit the House, then make haste to the same grid point, likely somewhere on the outer reaches of the local area, upon which the 'bleeped' tour would be closing in on, if not there already. At this improvised RV, the two tours would converge and brief. The standby team would relay the location of the TRA, the revised tasking, pass across a resupply of food, film and other comestibles and the two vehicles would part on their new missions, the standby likely completing the unfinished portion of the original's tour plan.

The 'bleeped' crew were now free to encroach into the newly restricted area, of course governed by normal rules of discretion and tactical cunning, and in apparent good faith could claim, if apprehended by the Sovs, that they knew nothing about its imposition. Depending on the activity that the tour detected, the original three-day plan could be extended by any number of additional days – family and personal plans had to be flexible. That was OP TALON SNATCH in operation – so simple yet so clever. As I said, I am not sure if the other two Missions employed a similar tactic to circumvent the restriction – certainly reading Thomas Wyckoff's book *Mission: A Cold War Remembrance* he seems to stress the point that USMLM sought to exploit the immediate period post-TRA lifting, when the exercising Soviet troops would have just left their training areas, to scavenge over the areas for documentary and material 'contraband' left behind, rather than any BRIXMIS-style sleight of hand – this was the reason Nick Nicholson was on tour that tragic day. Perhaps he was just being coy – I never had the conversation with American colleagues so can offer no informed opinion. However, I do not believe the Soviets ever figured out how we managed to frustrate and avoid the restrictions of TRAs so consistently. I regret that I was never on tour for a TALON SNATCH.

This was regular, 'normal' touring. I mentioned earlier the second broad type of tour: the local tour. It took its name from the fact that it remained limited geographically to the local area around Potsdam, a back yard of approximately 100 square kilometres. Given the density of units and activity in this relatively small region, local tours were fully justified in their limited geographic span. The local area was home to two divisions and all their subordinate units: 34 Guards Artillery Division (an honorific title dating from service in the 'Great Patriotic War') and 35 MRD. The 'local' also offered us the rail loading ramp at Dallgow-Döberitz, which was always a constant and fruitful scene for equipment sightings en route to, or returning from, exercise.

They were generally a less stressful operation because tours returned at the end of each day, until the tour was over, to the House for a good night's sleep and, more importantly, the superb fried breakfast prepared by the *Stasi* chefs working in the House's kitchen. I enjoyed local tours, not just because of the comfort or the shorter distances travelled, but for the intensity of reporting. There seemed to be a regular rhythm of activity and sightings whether it was hardware or soldiery or wives and children engaged in the normal routine of garrison life. The little kids in their Pioneer uniforms was always a conflicting sight – cute little things but a highly tangible testament to the nature of the Soviet system and its seeming future longevity. Even at this age the system was undiscerning, or perhaps highly discerning, in indoctrinating its citizens. Being able to observe married quarters where the families of officers and senior warrant officers lived was educational. They were generally dilapidated and in a very poor state of repair. But there was always something happening, always some activity to log and capture, without the *ennui* of the quiet days of a quiet tour. I also appreciated the realisation that one started to get to know the local area, around Potsdam, really quite well and enjoyed the sense of achievement in consolidating knowledge gained on previous local tours.

In parallel to the intensity and density of military targets, the local area was also disproportionately rich culturally and this certainly added spice and interest to a tour, in stark contrast to the bleakness of some of the towns, cities and areas we visited. A place I will never forget, for all the wrong reasons, is Bitterfelde, possessor of Europe's crown for most polluted town. It was without equivocation 'bitter' by name and 'bitter' by nature. There was a lingering brownness about the air, the buildings and the people, like looking through the lens of a camera fitted with a tobacco-coloured filter. The source of the pollution was principally the extensive mining for brown lignite coal – a thoroughly demoralising and depressing place.

In contrast, the centre of Potsdam was attractive in its own right, with better-than-average shops and a colour and character that was not typical of the DDR. There was the Dutch Quarter; the Rathaus, rebuilt after the end of the Second World War; and the neoclassical St Nicholas Church. On the outskirts of the ring to the east are the old Babelsberg UFA film studios dating from 1912, the oldest in the world. The west of the town hosts the imperial palaces of Neues Palais and Sanssouci and closer to 'home' the Cecilianhof, as I already mentioned. Sanssouci dates from 1745 and is a superb single-storey testament to rococo splendour. The gardens and surrounding park are equally impressive and, to their credit, the socialist regime had maintained both to a high standard. The New Palace, or Neues Palais, has much more of a British feel to it with its red bricks and copper dome. Then there were the thirteen traditional wooden Russian houses in the Russian colony built in the 1820s to house Russian singers attached to the Prussian army, and the attendant beautiful Alexander Nevsky Memorial Church. Further out lies the skeletal remains of the 1936 Olympic village in Döberitz – a fascinating historical 'remnant' sitting in forgotten fallow ruin.

By the time I had crossed the Bridge a few more times, some firm ideas relating to the nature of the DDR had begun to coalesce in my mind. Seeing this monolithic command state at close quarters, I began to understand some seemingly unassailable things about it. My *de visu* impressions were backed up by conversations wherever and with whomever I could have them, and by as much reading, both open source and classified, as I could access. I found it particularly interesting sharing thoughts and ideas with some of my USMLM colleagues and always enjoyed exchanges with my opposite number running their intelligence cell, Jim C, a US Army Military Intelligence lieutenant colonel – a really terrific guy who had family originally from the Baltic States. Attending joint ops or int meetings with the USMLM guys was always a positive experience, not only from the technical professional perspective but also linguistically. I have always thought Americans just have such great delivery and expressions that queue to roll effortlessly and colourfully off the tongue. It is entertaining and amusing just to listen to their turns of phrase. And not least, when they came to visit us in the Stadium, they would always arrive via Dunkin' Donuts and turn up with an armful of bags of sticky goodies – fantastic.

I somehow always experienced the feeling in our meetings and conversations that the Americans knew more and understood more than we did. Perhaps an element of this very un-British sentiment was based on the reality that their personnel shared wider cultural backgrounds that brought a more focussed understanding of the alien nature of DDR society; it was also undeniably explainable by the fact that, in contrast to the Brit policy of deploying the jolly amateur, the US posted in officers to USMLM with their more specific and specialist foreign area officer (FAO) professional backgrounds. They were certainly not all professional military intelligence officers, the majority coming from mixed all-arm backgrounds – infanteers, armour officers, engineers, gunners, even fast jet pilots et al. – but they were informed and, as Soviet-focussed FAOs, had all spent time behind the Iron Curtain as part of their training.*

Whatever the truths of this issue, it was apparent to me that the East German state certainly did work on at least one level. Regardless of the reality of the police state that clearly played a crucial and leading role in shaping societal matters, certain sectors functioned and worked efficiently and deceptively well, if 'well' is not too strong an adjective. The health system appeared to provide a good standard of accessible care. It was free, comprehensive and there did not appear to exist the plague of waiting lists. The education system, although highly politicised, provided learning to East German kids, preparing them for adulthood and, to those who qualified, places in industry and the apparatus of the state. There was full employment even though clearly many jobs were created and maintained with little connection to our Western ideals of supply and demand – that situation is in stark contrast to the high levels of unemployment that bedevils the post-reunification east now. And food and accommodation were heavily subsidised, ensuring that both essentials of life were affordable and adequate enough to maintain a level of fitness, health and protection. No one was homeless or living in abject poverty. No one was starving or malnourished. In contrast, I will never forget my first trip to Washington DC four years later and seeing homeless beggars on the streets a few blocks from the White House. That was a thought-provoking and sobering experience.

On another level, in parallel, it was, however, readily apparent that in terms of infrastructure, the system was close to collapse. Public utilities, including roads and motorways, appeared to be in such a poor state of repair, the result of inadequate investment, that I began to speculate about what was going to happen when things simply gave up. The state of urban affairs above ground was one thing; how were things below ground? I was certain that piping, cabling and sewers were just as close to collapse. What would happen when things stopped working and functioned no more or,

---

* All Russian-speaking tour officers were graduates of the Russian Institute at Garmisch in southern Bavaria. To illustrate how sought-after posts in USMLM were, of the twenty-five students annually, only three were selected to actually serve in the Mission.

at the very least, became so inefficient and ineffective that the daily lives of the citizenry was directly and adversely affected? How would the regime play this? What would the leadership and, as importantly, what would the people do, and how would they react? It was idle speculation on my part but it appeared to me that some sort of fundamental change of direction had to be on the horizon five, ten, fifteen years in the future. Of course my speculations were, in the end, overtaken by forces generated by other, largely external, factors, but it was intriguing to identify these huge question marks hanging over the star of the communist firmament. If the DDR was the wealthiest and best functioning of all the satellite states, then surely the future of Marx's dream was drifting steadily towards nightmare. These sound like ruminations fuelled by the benefit of hindsight but I clearly recall turning these questions and observations over in my mind as I watched the 'armoured cars and tanks and guns', to quote, completely out of context, the words of a famous Irish Republican song, and the conscripts of the Red Army training and drilling in seeming oblivious isolation, divorced from the political, economic and social realities beginning to swirl with increasing speed and momentum around them.

Since 1956, BRIXMIS had cleverly expanded its collection repertoire to include another form of very specialised but supremely effective tour. This was one of those in the 'other' category.

At the height of the Cold War all soldiers based within the British Army of the Rhine (BAOR), in West Germany had to be able to recognise at least the basic inventory of Soviet military equipment. Most photographs or slides of T-55 or T-62 tanks, BMP-1 or BTR-60 APCs or 2S1 or 2S3 self-propelled artillery pieces serving as source material would be marked with a lowly RESTRICTED. Occasionally the more astute would notice that a more interesting vehicle, seemingly photographed from atop a bridge or motorway flyover, would be classified as UK CONFIDENTIAL. This upgraded marking was not protecting a satellite image or the product of an off-course U-2 spy plane – such material would be considerably better protected. These images were the product of OP OBERON.

Unbeknown to its viewers, the photographs had been snapped over-wing, from the air, by a small RAF Chipmunk trainer. Nothing remarkable in this at first glance, except that, at the point of 'click', the Chipmunk was firmly, and legitimately, in Soviet-controlled East German air space. In fact, specifically within the Berlin air safety zone inside of a 20-mile radius measured from the Berlin air safety centre.

The more sensitive classification of the photograph protected the reality that the Chipmunk and its dedicated two-man crew, the pilot and an RAF SNCO in the front seat wielding the hand-held camera, were BRIXMIS members. Right at the inception of the Mission it had been agreed that the new quasi-diplomatic liaison team could retain its own trainer to allow RAF aircrew posted to the unit to maintain their flying hours. There were in actual fact two aircraft, allowing other RAF pilots stationed within the Berlin garrison on non-flying tours the ability to maintain professional currency.

Like every other function of the Mission's activity, the plane's utility was quickly morphed into a vital platform for intelligence gathering as the Cold War stand-off progressively cooled. A 40-minute Chipmunk flight could cover a significant part of the Soviet orbat, given the local area housed such a large proportion of GSFG/WGF's assets, and provide its unique intrinsic bird's-eye perspective of installations – an unimpeded glance at everything in the open within the tightly patrolled walls and fences – and equipment.

There was one final type of tour organised within the category of 'other': the cultural tour. They were mounted in stark contrast to the stresses, strains and frenetic action of 'normal' touring. On the scale of operational aggression, they were more related to the traditional liaison function, but no journey across the Glienicke Bridge ignored the ever-present imperative to collect any and all intelligence.

Cultural tours 'targeted' some of those jewels of Honecker's homeland – jewels then and jewels now – that really did reflect well, if not on the regime, then certainly on the country, on its heritage and history. These were places like Dresden with its art galleries, riverside location on the banks of the Elbe and the reconstructed beauty of the castle, and the Frauenkirche and Zwinger palace houses – places like nearby Meissen, the 'cradle of Saxony' with its rich heritage of porcelain production; Leipzig, the DDR's second largest city, famous for its trade and musical traditions, the showpiece fair and the Battle of the Nations monument celebrating the Prussian-led victory against Napoleon in the 1813 Battle of Leipzig.

Other destinations were less uplifting, and were, in reality, thoroughly depressing and dispiriting. I took part in wreath-laying ceremonies at a number of former concentration camps. Buchenwald, the largest camp within German borders, was established in 1937. It was sited on an open hilltop fully exposed to the prevailing wind. As we made the official tour prior to the commemorative ceremony, one could starkly appreciate the despair of the place and feel just how horrendous the lives of the inmates must have been in the middle of an icy East German winter. A group of us had travelled in full uniform using the Elephant Hotel in Weimar, incidentally allegedly one of Hitler's favourite hotels, as our base. It was an intensely moving experience. I also visited and paid tribute to the victims of the women's camp in Ravensbrück, and the KZ in Sachsenhausen. None of these camps were labelled as death camps, but the conditions and regimes imposed by the sadistic administrations running them directly caused the deaths of thousands upon thousands of prisoners rotting away within the barbed-wire perimeters. The communist-led regime, unsurprisingly, politicised the history of the camps, diverting blame and responsibility from the German people firmly and squarely onto the shoulders of the Nazi state. What they conveniently neglected to explain, and justify, however, was their continued use of a number of the camps for many years after the end of the war, well into the formative years of the communist SED regime, to house undesirables and enemies of *their* state – in reality, a strategy no different from the totalitarianism and inhumanity of their National Socialist NSDAP fathers and uncles.

The ceremonial aspect of the Mission's role was important and buttressed the other overt pillar of its role, its official liaison responsibilities. One of these ceremonial duties was to attend the large parade held annually on the square in front of the Soviet War Memorial in Tiergarten, perversely just inside the British zone on the western side of the Wall. The memorial, commemorating the 80,000 Red Army soldiers who died in the Battle for Berlin in 1945, had been constructed just months after the war ended and was made of stone taken symbolically from the destroyed Reich Chancellery. The site was permanently guarded by a Soviet honour guard that crossed from the east and was billeted in a small barracks attached to the memorial.

Personnel from all three Missions attended the parade along with a considerable Soviet presence including senior generals and diplomats from their embassy, just through the Brandenburg Gate on the famous Unter den Linden. It was an impressive turnout, well orchestrated and directed, and proved to be a moving experience as I stood to attention, proudly wearing my RCT No. 1 dress hat, looking very much like a bus driver, as the rousing and emotional Soviet, now Russian, national anthem was struck up by the Soviet military band. This is one of those totally rare and special occasions that I will, perversely, never forget.

On a historical theme I was also privileged to take part in a cultural tour of Colditz Castle, the former home of Oflag IV-C, the POW camp established to house the incorrigible repeat escapees of the Second World War, and made famous by the BBC TV show. The fortress perches, tautologically, up on its own hill above the small unremarkable town with its equally small winding river and railway station straddling the quiet rail spur. It was fascinating to have the opportunity to review the walls and towers of the castle and the courtyards so accurately represented by Stirling Castle in the television programme. Pre-Second World War, the site had been used as a psychiatric asylum, and on our visit it was apparent this more traditional role had resumed. It was off-putting and rather sad to see the faces of inmates pressed hard up against the windows as we inspected the castle – the modern-day prisoners. I could not help laughing though, overhearing the tutting conversation between two older women, outside the castle's fortified main gate, down in the town. They were expressing their lack of understanding, and some irritation, as to why we Western Allies would waste our time coming to their broken-down, irrelevant Colditz. The castle enjoyed absolutely no cultural notoriety in the minds or imaginations of the German population – our enthusiasm for the small town and its equally small *Schloss* was bewildering in their eyes. To them, we were making as completely random and pointless a trip as if we, a formal military delegation, had been visiting Huddersfield.

The aim of the cultural tour was two-fold. Predominant was the intent to 'fly the flag' as we had done visiting the former concentration camps as we paid official homage as representatives of our country – the UK – and our system – Western liberal democracy. We also sought to maximise contact with normal everyday citizens of the DDR as unofficial ambassadors of our world, our culture and our way of life. We wanted to clearly demonstrate that we were not capitalist monsters, we were not war mongers, rather we were ordinary people, like them, and we could laugh and smile and joke, and sit at pavement cafés and drink coffee and enjoy eating ice cream, not babies, just like them. In reality, our personal contacts were limited and embraced an incredibly small sample of the

DDR's total population, but we assumed that by word of mouth our reach would multiply naturally.

Additionally, these tours provided a release from the pressure of work and, for those who felt it, an escape from the confines of island-Berlin. Tours allowed wives and children to join their husbands, for those who were family men, and get a feel for their activities, or at least the environment within which they worked, and served as rare treats to witness first hand the realities, good and bad, of socialism at work.

There was a cadre within the Mission who rapaciously bought up some of the more 'commercial' items made in the DDR. This included porcelain, typically Meissen figurines and birds, model railway trains and other odds and ends. I am sure a proportion of these goods ended up for re-sale at some point in the future and I'm equally sure some of the ardent collectors made tidy profits. With East German marks exchanging at anything up to 20 to 1 *vis-à-vis* the capitalist Deutschmark, there were certainly deals to be had. I picked up very few souvenirs – it was always a case of putting off a purchase of this or a purchase of that: 'Oh, I'll get that next time.' And of course then suddenly it was all too late. I do remember, however, buying a pair of Zeiss NVA binoculars for the equivalent of £30 – they are great. I still have and use them.* And I did buy some Colditz porcelain just for the name, and some traditional blue and white patterned china; I still enjoy seeing the 'Made in GDR' logo on the base – history in the home.

I was much more into engaging with the conscripts that we ran into on the ground or visiting the 'chain' of DDR shops – sadly I do not remember the name – that traded in second-hand goods, a cross between a charity shop and a thrift shop. My interest here was to trade or buy the *matryoshka* dolls, quintessential symbols of Russian culture, if one could not afford or source Fabergé eggs or church icons, that the lads brought into the country from home and sold to make some pocket money. Cheap, low-quality examples predominated, but if you were lucky, you could always find better-quality ones, beautifully, but no doubt toxically, painted. For me, these were

* Gary had told me about his very first tour in the Mission. Another tour, coincidentally led by an Int Corps officer, had noted what appeared to be evidence of a sizeable delivery of Zeiss binos to their 'outlet' in Dresden. Gary called the shop from the ops room, and on hearing the positive news, raced down to Dresden, himself and a driver, and bought up the entire consignment. Everyone in the Mission who wanted a pair was able to purchase one. Like me, both Gary and his dad still use them.

mementos that all told a tale and were more representative of the unique reality we confronted as the Military Mission.

Cultural tours were fun. They were educational, providing an unparalleled view 'through the looking glass' into an alien but very real civilisation, and they were the catalysts of the lasting, broader memories of a country that was to disappear virtually overnight, leaving almost no trace of its transient former character and identity.

10

My Touring

What motivated me about touring in general was fundamentally this opportunity that it provided to observe, witness and study the political philosophy in action that we regarded as the most physically dangerous and morally corrosive enemy that we, as a military – and we, as a state, society and member of a wider alliance – have ever confronted. I never tired of being inside the heart of the enemy camp and literally rubbing up against its people and its institutions. Every day over the Wall generated impressions and memories from the bleakest to the lightest. I could not resist smiling observing on a number of occasions tiny *kindergarten*-aged kids being towed by their teachers to or from nursery school in small trolleys/trailers. And what about tuned up Trabis being raced on frozen lakes, their tyres studded to provide grip?

On another level, as I shared the pervading doubt that the Warsaw Pact would ever move preemptively against the Western alliance (nor for that matter that NATO would ever launch a first strike the other way), I was not fired up, as I explained, by the drive to tap into the drip feed of low-level intelligence that confirmed normality, but I firmly do acknowledge that it was undoubtedly important as a confidence-building, and maintenance, measure. To reiterate, what I considered most important was the collection of technical intelligence ensuring that a balance of capability was maintained that dissuaded any idea to exploit a perception of weakness and make a quick strike accordingly – no more bomber gaps, missile gaps or any other gap. But the thrill and excitement I definitely got from the job was just to see at close quarters the tanks and APCs, the SP guns and all the other wheeled and tracked alphabet soup, and the foot soldiers of the Red Army at work and play. For me the thrill of catching a column of T-80s with the distinctive scream of their gas turbine engines bouncing along the rough of a well-used tac route, hull up and hull down like a small ship on a stormy sea, in the middle of nowhere, was beyond measure. The thrill was entirely one of face value. It was what they were and what they represented as powerful impressive machines and servants of war, rather than an element of such and such a battalion of such and such a regiment of such and such a division, that motivated me – frankly, as I've said already, I had little interest in that. What I valued was the impression of a company of motor riflemen, infantry soldiers, clearly comprising representatives of a cross-section of most of the diverse Soviet Republics out on the road running in squad formation in the same uniforms they would obviously be wearing for the rest of the day and quite likely the rest of that week, unchanged. Watching them was seeing the face of our enemy, and instead of further demonising him, it served, rather perversely, to humanise him and actually to continue to reinforce the process, that our privileged position provided visibility of, of breaking down the image of the invincible Soviet horde that seemed to pre-dominate our conventional wisdom and popular assessment. He, the Red Army conscript, was a member of a formidable fighting force but one beset by as many weaknesses as strengths. Lessening and controlling fear reduces tension, which reduces threat, which enhances the prospect of enduring peace.

What then do I remember of my times on tour? Rather than recalling specific memorable tours from start to finish as complete operations, I have a mosaic of disparate and largely disconnected experiences and images in my head – I guess, unless it is photographic, that's how memory works anyway.

The operational aim of every tourer was to photograph a new item of equipment in the Soviet arsenal, or at least in the GSFG/WGF orbit. Because of its front-line status, GSFG/WGF attracted the newest and most effective equipment. During the tiny window of 1986–89 more new weapons systems arrived to equip the five Soviet land armies, and massed divisions of the East German NVA, than at any comparable period during the whole Cold War. As I already mentioned, my colleague and friend Steve

Gibson scooped the first SA-10 in the DDR, with its distinctive four large missile pods, and earned an MBE in the process.*

There had been talk amongst the tourers of an unidentified vehicle. There had been several inadequate long-distance partial sightings that had been sufficient to suggest something new and different, but not specific enough to put a face or name to the stranger. It had been seen nowhere else in the country except on this one particular small training area on the outskirts of Brandenburg, 80km west of Berlin. It was consequently unofficially christened 'The Brandenburg Beast' as a prelude to any official NATO designation.

It was my luck to be right place, right time and to have spotted the Beast racing around a driver training circuit out on the training area. We made the decision to approach boldly for a quick one-time assault before lifting off and moving briskly out of the immediate area. Exploiting our similar profile to the Soviet UAZ-469, we were able to approach the circuit without challenge. From the back of the G-Wagen, I fired off a salvo of pics capturing the Beast and slaying the myth before we demonstratively left the scene.

Sadly, the reality was disappointing and completely failed to capture the attention of the technical intelligence audience, who had held their collective breath for Steve's scoop. What we had photographed appeared to be a local modification to a MT-LB chassis, basically an artillery tractor/utility vehicle. A popemobile-style box had been fixed to the top of the vehicle and a driving position created, transferring control from the recessed location that was a standard feature, tank-like, at the nose of the machine. The trainee driver sat atop the Beast and consolidated his driving skills with the enhanced visibility that the model provided. We might have failed to shake the earth, let alone the Olympic Stadium, but at least we had put an end to the speculation about the role and purpose of this ersatz vehicle.

My most impressive and lasting image of touring took place one dull autumn day. We were on the outskirts of Schwerin in the north of the country up towards the Baltic coast. We were travelling on the main road en route from somewhere to somewhere else. All was quiet if not to say

* The SA-10, or S-300, was to feature prominently a few years later when I commanded a unit in Cyprus. The Cypriot government somewhat provocatively, but entirely within their right as a sovereign government, purchased a number of systems from the Russians, thereby giving them the ability to shoot down Turkish jets well inside Turkish airspace and readjusting the regional balance of power.

normal, if not to infer rather dull. It was afternoon, and suddenly from right to left, maybe 800m to a kilometre away, the sky seemed to fill with Mil Mi-24 'Hind' gunships. These are the intimidating insect-like attack helicopters so often filmed on mission during the Soviet misadventure in Afghanistan. I counted fourteen Hinds in total: a sinister, evil-looking swarm of hornets roused from their nest, angrily seeking the intruder. It was an impressive and powerful sight. I still consider the Hind to be the most aggressive-looking military machine designed by man. I was so glad I was not one of those *mujahidin* who had to face them in battle, even if I had had a Stinger launcher sitting comfortably on my shoulder. In intelligence terms this sighting signified nothing and would have been chalked up as a non-event, but it made me happy.

We had a Royal Navy warrant officer in the Mission whose role was to report on seaborne affairs along the stretch of Baltic coast between the old Hanseatic city of Hamburg in the west, and east to the Polish border. Despite this, I was tasked on one of my tours to take a run along a section of the shoreline and see what was happening nautically up there. It was, of course, midwinter and the weather was cold, wet and foggy, like a scene from *Riddle of the Sands*. Visiting this part of the coast in summer was a very different proposition and is actually a well-kept tourist secret. But now I just wanted to get things done, leave this dispiriting mean-looking stretch of dunes and grey rollers and move on to my main task on the island of Rügen to the east. We were able to photograph a couple of East German frigates without interference and accelerated away through the mist and driving rain towards Stralsund, happy to be leaving this desolate and energy-sapping place.

Stralsund was noteworthy because of its role as a point into and out of the DDR for the movement of ship-borne equipment. It's highly likely the SA-10 that Steve had scooped had begun its operational tour on this route. We manoeuvred our way to the best OP site as detailed in the target pack back in the Mission without drawing any undue attention, hoping we might get a sighting of noteworthy activity inside the port complex. The fact that things were quiet in the surrounding area and along the main access route suggested the docks would be equally '*Toten Hosen*', as the Germans say.*

* A great analogy meaning literally 'dead trousers' – superlatively quiet. Also an excellent new-wave/punk band from Düsseldorf.

As a good rule of thumb, an early indicator of activity or imminent road movement was the pre-positioning of 'regulators' or 'reggies' as the Missions referred affectionately to them. They were essentially human signposts, a role undertaken by the British Army's RMP, whose task was to cover a significant road junction and ensure that convoy members took the correct fork. They could spend days at their lonely post and were usually the subject of a quick stop by tours for a chat – always, of course, including an enquiry as to what was going on – and a gifting of western cigarettes or chocolate. I always felt so sorry for them marooned in often awful weather, numbed by cold and supreme boredom and sometimes very poorly provided for, food and equipment-wise. At the height of summer, the level of discomfort was equally extreme, with little water and less shade. Better though perhaps than a tour in the mountains of Afghanistan.

So no reggies, no action. And so it turned out, but another target visited and negative intelligence for my report. The production of reporting that attested to a continuance of the status quo was just as valuable to the planners 600km away back in NATO HQ in Brussels. This still represented success for the Missions and was an integral part of their utility as a strategic reconnaissance asset.

We then crossed over the narrow channel of water between the mainland and the island of Rügen. There were targets of pressing value here but this little-known jewel possessed a fascinating history of its own. It was the site of the amazing Nazi-built 'Colossus of Prora'. Constructed between 1936 and 1939, it was part of Hitler's vision to build a massive holiday complex providing affordable holidays for German workers who would be able to enjoy the long beachline and sun and sea during the long months of summer. The camp consisted of eight identical blocks stretching 4.5km from end to end, and was an impressive monument. It is fascinating psychologically that totalitarianism feels impelled to build these monstrous edifices, but perhaps ultimately no different from any powerful historical force. The island supported the Mukran ferry, another maritime route into the DDR from the port of Klaipeda in the then Soviet Baltic Republic of Lithuania. No reggies again and more dead trousers.

Another snapshot still brings a knot to the pit of my stomach. I was on tour with Ed, himself a part-time tourer as the officer running air operations. He was a squadron leader; as a seasoned pilot he had been the chief instructor at the Central Flying School and was a lovely guy and tremendous company – we may not have been collecting priceless intelligence for the community but we were having a good time.

We had pulled up at one of the numerous large railway sidings peppered throughout the country and were cautiously navigating our way into points of vantage so we could record what equipment was temporarily halted on its journey from garrison to training area or vice versa. The key was to spot the points on the trains where military guards were protecting their equipment wards and from whence they could spot us and potentially take action. Always at the back of a tourer's mind was that an edgy, even potentially drunk, Soviet sentry could over-excitedly take a dislike to us and shoot. History was always capable of repeating itself and periodically did so.

However, equipment temporarily unguarded and resting in these situations presented unparalleled opportunities, security considerations duly factored into our approach, for us to get in close and either record comprehensive details, check routing information on the rail dockets or potentially even more. Recorded in the annals of the mythology of touring, the calibre of the main armament of a new armoured vehicle had been confirmed by a switched-on tour NCO spotting a trainload of them on flatbeds, in just such this sort of situation. Seizing the initiative and thinking on his feet, he had pushed his uneaten lunchtime apple into the muzzle of its main gun to record the impression. One version asserts that the main gun, 125mm, was on the new T-80 tank. The other fable records the new vehicle as the BMP-2. If the fruit in question was indeed an ordinary NAAFI apple, I tend to believe that – unless it was a monstrously large Granny Smith, more the size of a grapefruit – he was tackling the BMP tracked armoured personnel carrier with its 30mm auto cannon. Back in Berlin it had been a simple matter to convert the fruity indentation into a measurement that could start to guide the calibration of the west's technical countermeasures.

On another occasion, another tale retells the story that, with the help of some simple tools from the back of the G-Wagen – and no doubt every soldier's friend, the Leatherman – a plate of new explosive reactive armour (ERA) had been successfully liberated from the front of the hull of a T-80, again in transit, that had halted in a railway siding. (A conflicting version recounts events in a slightly different way, crediting this scoop to a speculative rummage on a tank firing range stumbling across an object partially buried in the loose, sandy soil. It was quickly ascertained that the green, hard oblong box was ERA. No one was more surprised than the inquisitive digger. Whatever the veracity of the circumstances – and I suspect the relative truth matters only to the storyteller – it was easily smuggled back over the Bridge to Berlin.) The ERA box had revealed that the level of protection this new tank armour afforded totally negated the effectiveness of NATO's shoulder-fired anti-tank weapons. This was technical, strategic-level intelligence of the highest value.

This particular siding that we were cautiously feeling our way around was so packed with normal civilian rolling stock that we were unsighted as to what lay concealed in the middle of the silent, unmoving mass of railcars and flatbeds. Deciding we needed a vantage point from which to survey the site, I climbed out of the vehicle and made to climb onto the bonnet of the truck secured on one of the flatbeds nearest to our open track. I double-checked there were no sentries, or for that matter *Deutsche Reichsbahn* rail workers, watching us.

It was one of those moments, as I negotiated the front bumper and found my footing on the smooth, rounded, khaki-painted metal of the bonnet, when I intuitively sensed the warning that something was wrong. Something felt wrong. As the ghosts of all my ancestors pointed furiously in a strangled mime desperately trying to pass on their warning across the boundaries of time, I paused. Even in my tired, fatigued state after two long days already out on tour, including a TOMAHAWK, I now knew something was not right; there was an, as yet, unidentified threat lurking, about to strike, and I had to exercise some sort of existential caution, now.

I looked up just as I was about to stand and assume my full height on the bonnet of the truck to look beyond and see what I could see. What I could see instantly sent a bolt of super-charged adrenalin coursing around my system, jolting me into that instant state of hyper-alertness, energised for emergency and action. I had been just about to raise myself up into the overhead high-voltage cables strung above this section of line. Chilling. I was literally a few inches from instant death as, unconsciously, my head would have made contact and completed the circuit, conducting however many thousands of volts into my body. Needless to say, I remained frozen in my semi-crouched hunch. But I did have a good look, confirmed there was nothing of military interest to my right, parked up along and across the tracks and proceeded to earth myself firmly back on hard terra firma. Neither Ed nor the driver had noticed my 'near-death experience' and I chose not to share it with them. An interesting idle thought though: what was the Mission SOP for handling a fatality whilst out on tour? Actually not such irrelevant speculation as both the Americans and the French had lost tourers out on the ground and, but for the grace of God, there could well have been more.

The French fatality had occurred when a senior warrant officer, Philippe Mariotti, was killed in a RTA near Halle on 22 March 1984, when his tour vehicle was deliberately rammed by an NVA truck. His front-seat fellow tourer survived but was still receiving medical care twelve months later. In a more sinister incident, as I have already described, Major Nicholson had been shot dead a year later on the outskirts of Ludwigslust, and in late 1987, a SNCO, Charles Barry, also from the American Mission, was shot in the arm whilst on an air tour but had survived his wound.*

More snapshots. We had spotted a concentration of troops in a wood. They were all dressed in full nuclear biological and chemical (NBC) warfare suits. The Soviet pattern, then, comprised a green/grey impermeable rubber-coated canvas over-cape, hooded to enclose the GP-5 '*Schlem*' respirator, giving the troops a decidedly, almost comic, pig-like appearance. Despite this, the whole ensemble was nicknamed the 'Womble' in tribute to the British TV programme dating from the late 1960s. Anything relating to the whole realm of NBC was high priority as a collection target.

We moved slowly but purposefully closer as the driver manoeuvred the vehicle at low revs. Exchanges within the vehicle, in fact both on tour and back on the corridor at the Stadium, were very informal following the precedent set by all Special Duties units – no 'Yes Sir, no Sir', no 'Sergeant this' or 'Colour Sergeant that', rather exchanges based on 'Boss', first names or nicknames at working level. But at these times, as we moved into target-tackling mode, the whole tone and flow of conversation within the vehicle altered radically and changed a gear. No more banter, no more humour,

---

* Arthur Nicholson was promoted posthumously to lieutenant colonel. I always thought this was a fitting gesture made by the US military that would not have been mirrored by our own government – it is a sad statement that the case for a bar to the GSM to be awarded to Mission personnel for their not-insignificant contribution to Cold War victory, and in recognition of their supreme professionalism and personal courage, a tiny gesture, was unceremoniously rejected. But the example of the US Defense Department serves as a solid statement recognising his personal sacrifice, emphasising and reiterating the US government's appreciation of Lieutenant Colonel Nicholson's professionalism, dedication and courage as a soldier, and as a very practical gesture it would have increased the size of the widow's pension that his wife would receive for the rest of her life. Not generally seeing a value in reunions, I met Nick's wife and daughter during my chance, unplanned attendance at a Mission one in Berlin in 2002.

no more gossip or irreverent speculation. The crew flicked instantly into operational mode, and communication assumed the crisp, curt, measured delivery that the military excels at. Directions to the driver were clipped and unambiguous, directing him on speed, direction and outlining contingencies should the situation escalate. Dialogue between tour officer, the commander and the tour NCO focussed on the target and the approach to capture the maximum amount of information. This raw information became intelligence once we returned home and the various Mission offices processed it, analysing it, collating it, assessing it. No word was spare; no word was wasted or redundant. The key was clarity and brevity – more basic military principle and lessons.

We had still not been seen but it was now just a matter of time – we were literally almost at an arm's length from the 'piglets' as they exercised just inside the wood-line. This was a situation waiting to unfold, equidistant from either an attempted controlled and coordinated detention, or an accident, if a Soviet decided to launch himself deliberately and recklessly at the vehicle. I told the driver what I wanted to do and how I wanted to tackle the target, such as it was. Sometimes I could feel the connection to my Highland heritage, and when we no longer required the guile of the ghillie, it was good to charge the enemy. So we picked up speed and did a run along the treeline, photographing the startled Wombles as we raked them with camera fire. Like a ground-attack aircraft strafing a target, we pulled off at the end of our run, swinging right towards the main road, job done, and moved smartly out of the immediate area.

Again, we had discovered nothing new and nothing worthy of update but, equally, by recording what we saw, we confirmed that we knew what we knew – not really negative intelligence but not really positive either.

I was pleased, in order to complete the set, to deploy on one air tour. We targeted Finsterwalde, sitting midway between Berlin and Dresden, a fighter base from which Sukhoi Su-24 Fencer nuclear-capable fighter bombers were active that day. The Fencer was the Soviet answer to the General Dynamics F-111, mirroring its variable-sweep wing and 'side-by-side' two-man crew.

We were aware of the orientation of the main runway from the target pack and knew how to get ourselves into a good OP position on the centreline to catch the jets recovering at the end of their training sortie. All G-Wagens had a roof hatch, a circular mini turret that, once opened, allowed the tour officer to stand on his recessed seat and operate camera gear shooting over the vehicle's roof. It was a perfect way to tackle airborne targets from the security of the vehicle. If things became hot, it was a simple matter of dropping down inside the wagon, closing the 'lid', submarine-like, as Jesse Schatz had done on that fateful, tragic day, at the same time as the driver gunned the engine for a swift getaway.

We did not have long to wait before the returning Fencers 'glided' in, approaching the runway threshold on their final descent back to their hardened shelters. It was a truly amazing experience as these front-line fighters seemingly floated over our heads barely metres above us. The noise from the twin jet engines was incredible. We could see every rivet, every bolt and, of course, every underslung munition. I felt the crews must have noted us as clearly as we saw them but for whatever reason they seemed not to have alerted airfield security and we watched the whole show without interruption, whilst I photographed each and every jet. It was spooky to see the red stars on the tail fins at such close quarters – but for these, it could have been Ashford training up on the East Anglian flatlands – as the tour NCO recorded onto his tape the bort numbers identifying each jet.

Of course, predictably nothing out of the ordinary to report and certainly no underslung nuclear weapons.

On a bright, sunny afternoon we had been tracking across country en route to check a small training area highlighted in the tour plan. It was the second day of a standard three-day tour and we had encountered nothing beyond the routine. As we swung around the apex of the bend in the minor road we were now traversing, there, set back hard into the treeline in front of us on our right was a distinctive Snow Drift mobile target acquisition radar. Just as rapidly and instinctively as I knew that seven times eight was fifty-six, I knew that, supporting the SA-11 Gadfly, the radar boasted an impressive technical spec. It could see out to 85km and detect low-flying air targets at 100m at 35km, that the SA-11 missile would then turn into so many scorched metal fragments. The radar was mounted on its own tracked chassis so was fully mobile. Whilst it was a relatively new piece of equipment to GSFG/WGF, it had been photographed before and the Allied knowledge of its capabilities and tactical deployment was well known. Nevertheless, its presence, sitting unprotected in splendid isolation, presented the perfect opportunity to update and add to the existing photographic catalogue. Regardless of whether we assessed this was old news, we were always looking to identify modifications and design tweaks that would enhance capability and performance. It had to be photographed so the technical experts could officially confirm that the threat it potentially posed was correctly assessed and unchanged.

We came to an abrupt halt and moved off the road into the cover provided by the wood-line of the copse on our left. The deciduous trees were healthy, the branches low and fully in leaf. Man and state might have been against us, but nature in the DDR was invariably on our side and worked tacitly to aid our endeavours. I always felt comfortable and optimistic when out and about in the countryside of the east. There was a vitality and an enduring robustness and healthiness about the land and its flora and fauna. That this was the consequence of SED policy, I doubted. More likely it was an unintentional testament to the state of the east's economy, its backwardness and undeveloped nature, particularly rurally compared to the intensive and aggressive nature of farming in the west. A valid comparison could almost have been the pre-enclosure nature of Britain's landscape contrasted to the violent changes that more intensive and radical farming methods wrought on the land with the coming of the agricultural revolution. Whatever, it worked on this one level, even if the regime struggled to feed its own people without aid from Mother Russia.

We were now a good 400m away and essentially invisible, the dark green of the Merc blending seamlessly with the browns and greens of the trees standing sharply to attention, closely ranked behind us. Following the lessons of basic infantry fieldcraft, we were always conscious of the implications of 'shape, shine and shadow' and were positioned so the gorgeous sunlight was reflecting brightly and beautifully off other surfaces but not our own windscreen or windows. I reached for my binos and swept the target. I could clearly see a couple of Sovs moving around the vehicle, but they had not seen us and were intent on their own business. I then raked the surrounding area to check the Snow Drift was not part of a larger concentration of equipment that we had thus far failed to spot. It was alone. I returned my focus to the radar. It was in the raised position and it looked like they were in mid-exercise mode. It was apparent that, unless things changed rapidly, we had the time and opportunity to take some leisurely high-definition pictures. The photographic light was absolutely perfect and could not be bettered. My focus switched to the equipment at my feet.

I moved to change lenses and reached into my large, hard canvas photo 'box' where I stored my additional two camera bodies and large lenses. I replaced the standard 85mm lens that served as the default 'window' and reached for the large-diameter chunky 1,000mm mirror lens and the small doubler – as a point of reference, the human eye registers images equivalent to a 50mm lens. I attached both to the Nikon F3 body, transforming its relatively sleek profile into a meaningful-looking piece of serious hi-tech machinery. The F3 was a great camera, reliable, easy to use and robust, constructed from titanium in the version that we deployed. It was the body of choice deployed throughout the armed forces and I could use it to best advantage blindfolded, upside down on a roller coaster. These were still the days of Nikon's unchallenged supremacy in the 35mm SLR stakes before Canon changed the game with its gyro-stabilised lenses.*

The 1,000mm lens – effectively now a 2,000mm monster – dramatically brought the picture taker right up close to the subject, to within hand-shaking distance. Doubled, we were now eyeball to eyeball. It was a great lens for wildlife photography and a great lens for our brand of espionage work. The disadvantage was that it was heavy and, unless really sturdily supported, could only be used at fast shutter speeds to prevent the inevitable camera shake blurring the image. This meant that in the trade-off that underpins the art of photography, the constant balance between light and speed, it was impossible to maintain any degree of depth of field (simply put, the degree of sharp focus maintained before and behind the photographic subject); the aperture setting was too low. Having effectively transformed the lens into the giant it had become, I had simultaneously enhanced the detail I would capture but made my job even more tricky in the process. However, with the light on my side and sitting inside the G-Wagen, I actually had a less daunting and relatively straightforward task. I recovered my beanbag from the box and cranked down the window. Mirror lenses could not be used shooting through glass, as the resultant reflection produced blurred, sub-standard results. I rested the lens securely and solidly on the bag and window frame to produce the sturdiest platform possible – better than any unwieldy tri- or bi-pod. Now I was able to conquer shake whilst

* We received some trial F4s from Special, a bigger, bulkier but updated camera with integrated auto-drive. I immediately fell in love with it – so much so that I bought myself one, and they were not cheap. Consigned now to storage with my other wet film bodies, I am still to some extent a reluctant forced convert to digital technology. It has been an expensive transformation.

extending depth of focus, a perfect low-tech, no-nonsense solution – my favourite approach to any practical problem.

I took a couple of rolls of film, zeroing in on every feature of the business end of the radar, its support and the chassis. I felt a bit like Rankin at a fashion shoot as this Russian beauty pouted back at me. As I clicked away methodically, adjusting and re-adjusting the manual focus, as I subtly moved the lens over the image, the driver and tour NCO kept their eyes peeled, ensuring we were not taken by surprise by the 'wandering Sov' stumbling across us in our enhanced state of vulnerability, my window wound down and ogre-like lens protruding like a boat race oar.

Once back in the Mission, I studied the prints that Special had developed. They were absolutely perfect and some of the best I have ever taken – crisp, sharp focus, beautiful warmth and popping, eye-watering detail. Coffee table book stuff if you loved Soviet air defence search radars.

But, again, you've guessed it – nothing new.

We had a discreet scout around but were not able to identify any other components of the SA-11 system. It looked like the radar boys had been exercising out in the field independently, with perhaps everything else happening back in barracks. We moved on.

I was excited to tour out to the far east, taking in the barracks of the airborne regiment, with its distinctive finely hooped blue and white T-shirts, en route through Cottbus – nothing significant to report. As we hit the extreme eastern edge of the DDR, having passed through Frankfurt an der Oder – a main rail route into the country from points east, and definitely not to be confused with the banking centre – it was something special, to me, to stare across the frontier, over the narrow waterway of the River Oder into Poland. The simplicity of the red and white stone border markers somehow sharpened the impact of the experience. Here I was, standing firmly on communist soil, looking even further into the empire of evil, into historic, and blighted, Poland. I could see a farmer toiling in the distance, an old tractor, but no other technology, resting nearby. Otherwise all was quiet. I could certainly have been looking backwards in time three or four generations. The stark realisation, as I stood on the banks of the river, was that Europe – my continent, all the countries within it that I had visited and passed through, so familiar, warm, welcoming, dynamic and forward looking – was in reality two very separate continental plates. It was false, deceiving, simply inaccurate to talk about 'Europe'. This eastern part was markedly and fundamentally different. The reality was that there was West Europe and East Europe, just as there was North America and South America. As I watched, despite my questions about how the SED regime, if not others within the bloc, would deal with the social problems and economic issues I was sure loomed ahead, I felt there was a sense of immovable, unshakeable permanence. Surely nothing was fundamentally going to happen to change things even if I was to look forward three or four generations. Somehow this sense of permanence was reinforced by the simple coloured border stones. In their understated way, they projected an image of continuity reaching backwards and forwards in time, like the stone circles of Orkney, suggesting they would remain long after I had left the DDR and, for that matter, the Army too. It was a thought-provoking, but certainly not an uplifting, depressive experience.

As I turned back to the G-Wagen, I remember a lighter more frivolous thought: what highlights were my friends from university facing that day? I suspected things more mundane, but perhaps, hopefully, less crushingly depressing – probably though considerably better remunerated. As we pulled away, back towards the lights of home and Berlin, I again thanked my stars for the rich experiences that still seemed to be blessing my life.

Once again, nothing to note beyond the experience itself.

Another thought beyond these began to form in my consciousness and needle me. Had I already lived the most exciting and challenging days of my career? Frankly, the Mission was fun; it unquestionably presented incredible personal opportunities, memories and images, and a professional tick in the box that only added positively to my CV – all things I would treasure and never forget – but as an operational intelligence collector, I felt I was coasting. I thought back to the pace, pressures and challenges of FRU days and could not help but feel a sense of professional anticlimax. I thought about the ever-changing diet of intelligence that we collected from our agents against the operational background of a bloody counter-terrorist war and compared it to the trickle of my sightings and photographs that had invariably already been recorded dozens of times before, 'Brandenburg Beast' exempted. I knew that every tour was just a moment away from a significant scoop but, ultimately, this would serve only as a blip on the wider screen. The basic currency of the day-to-day collection effort was distinctly nickel, in my mind, far away from the silver and occasional gold of agent operations. I recognised in myself that this was a negative, corrosive and fundamentally unrealistic attitude, that life could not function on a permanent high, travelling at Mach 2, but equally I knew the more you have, the more you want. However, little did I know that the world was soon to flip so dramatically that everything, including my doubts, was soon to be totally upended, and we were to experience the most far-reaching revolution since 1917.

It was not uncommon for the Missions to deploy on tour in response to 'tac tip-offs' from the strategic intelligence agencies. These were essentially prompts that steered the Mission teams to areas where the reporting of activity required further corroboration and detail. This was usually because the reporting was ambiguous, unclear or incomplete and no firm analysis could be conducted. The most prolific steerage emanated from 'The Hill', the SIGINT squadron based in Berlin on the Teufelsberg. There was no better or more responsive collection asset than the Mission, capable of providing this *de visu* reassurance in this context.

A tour would cross the Bridge and, conscious of acting with the required discretion to protect the sensitive source of their tasking, move to the location or area highlighted by the signals intelligence and physically record the activity, the essence of a perfect symbiotic partnership. Reacting to a tac tip-off, Gary, supported by the then Regiment tour NCO, a former Int Corps corporal who had passed SAS selection, had come close to scooping the SA-10. They had caught the top of the Flap Lid radar from extreme range, shooting into a PRA from its boundary with the camera and long lens mounted on a tripod on the G-Wagen's roof. Their intelligence was enough to confirm that something new was in-country, the spur to push other tours, from all the Missions, to increase vigilance and output.

However, significantly from my own personal professional intelligence perspective, I was tasked by Ops to flesh out intercepts on two occasions. In both instances, having toured to the area highlighted, there was nothing to report, literally nothing. The area was clear of any Soviet or East German military activity. For me, it was educational to demonstrate the fallibility of an agency that invariably reported intelligence with such self-confidence and certainty. My individual experiences certainly raised some important questions in my mind. It also definitely underlined, again, the utility and value of HUMINT, information provided, in this scenario, by *de visu* sightings from reconnaissance missions, definitely a sub-strata of human intelligence. I knew, in the past, similar tasking had revealed systematic and professional disinformation operations mounted to confuse and mislead

Allied SIGINT interception. The Soviets were experts in this and laid great store on the art of deception in all areas of endeavour against their enemies.

The added value that the BRIXMIS product generated was that, in these instances when the Mission reacted to steerage from the strategic 'big boys', and activity was confirmed and definitively identified, it enabled the product to be disseminated, rather than with the normally heavily restricted TOP SECRET classification to protect the highly sensitive source, at the lower, usable CONFIDENTIAL level. This was the protective caveat under which BRIXMIS routinely reported tour intelligence. This ensured intelligence was more likely to reach users, sitting much further down the food chain, who could exploit it. This was a significant and important factor that endured throughout the Mission's working life. The intelligence that tours collected could be used at the tactical level where it might make a difference. The whole rationale of collecting military-related intelligence is to use it to inform operational planning. If the circle of potential 'readers' is so restrictive, the tax payer cannot be blamed for scratching their head in perplexity. There were BRIXMIS reports that were more highly classified, but these related to very specialist, and relatively infrequent, tasks that the Mission undertook.

Recalling my touring highlights, I also, excitingly, got to 'run' a Soviet column. This was the potentially highly hazardous and stressful process of 'attacking' a moving column of vehicles out on the open road. There were two approaches: from the 'front' and from 'behind', no other allusions intended. From the front saw both Mission vehicle and column passing each other from opposite directions, like Red Arrow singletons. Of the two, from behind was exponentially more dangerous, as we ran into the rear of the convoy and then progressively picked off the constituent vehicles and 'packets' as we overtook them and jockeyed back into the gaps. Incidentally, this was illegal under East German law, if one acknowledged the legitimacy of the state, which we did not. This approach was likely to produce better and more comprehensive information. The danger could come from two different directions, literally. There was the 'normal' element inherent in any overtaking manoeuvre – the risks presented by oncoming traffic, poor road conditions, manoeuvring space and just plain sod's law. Fortunately for us, on this occasion, it was night, and the dark ensured that oncoming road users were more easily spotted as they should but terrifyingly might not be using headlights. And the snaking configuration of the column gave the driver a feel for bends and other hazards ahead.

The dangers posed by the column itself were altogether different. Once the Mission vehicle had been identified, there was the real danger of an over-zealous conscript driver attempting to maintain the integrity of his column by closing down gaps into which we could slip, or aggressively attempting to side-swipe us and run us off the road. When there were radio comms between vehicles, this threat escalated dramatically, as a warning of our intent could be transmitted up the convoy's length for other vigilantes to watch for us and take action. Some convoys could caterpillar for several kilometres without significant break, so the dangers racked up and the odds of confronting a dangerous situation multiplied significantly.

We skilfully, purposely and with controlled aggression of our own, started to move up inside the column, passing vehicles and pulling in when safety dictated it. As we passed each tailgate, my tour NCO called out the type, the VRN on the number plate, anything visible in the back of the trucks, and any other pertinent detail into his mini tape recorder. From my back seat I photographed the registration numbers for the Research boys, providing the mandatory Mission 'stereo' effect, and was ready to record any significant kit more carefully. Things got pretty steamy and hectic as, after every thirty-eight vehicles, I had, of course, to change film. The reality was not quite so starkly furious as, with my three bodies, it was more a matter of handing the full camera to the NCO in the front passenger seat, who would wind in the exposed film and snap a new one in place, multi-tasking as he talked to the tape but still within his capability limits. Once I'd confirmed the camera settings, I could then recommence shooting. In this way, it was possible to capture close to 100 per cent of the vehicles making up the column. As we passed we would check the driver's cab to confirm the driver's arm of service.

On this particular night we experienced no problems and no additional hazards, as we apparently were not recognised as a Mission vehicle in a timely enough manner, or the column was comms-less. It was always to our advantage, especially under cover of darkness, that we casually resembled the UAZ-469. The column turned out to be a standalone one and relatively short – good experience and a definitive change of pace measured against the routine activity of the day. Back in the office, the VRNs were processed and the convoy recorded and its relative (in)significance assessed.

In terms of any significant intelligence: nothing, again, except perhaps to add weight to my doubts about the utility of what we had done. In terms of balancing the two key operational variables of risk and gain, was what we had performed, the potential dangers we had faced, worth the result? Because if my analysis and conclusion were valid, all we had confirmed was that so many trucks of whatever unit from wherever they were based had been out for a drive, purpose unknown, at such and such a time, in such and such a direction, with the classic concluding comment so often seen in operational intelligence reporting: NFTR – Nothing Further to Report.

A highlight, more negatively shaded, at least in the eyes of some tourers, engulfed me and the team as we headed towards the beautiful market square, cautiously and with our usual restraint, through the centre of Halle, birthplace of Handel; the infamous Nazi Reinhard, Heydrich; and former German foreign minister Hans-Dietrich Genscher.*

We were consciously refraining from taking any undue attention of anything on the city streets, except the summer-clad local girls and the ice cream and bakers' shops where we were considering stopping to make a break and vary the reality of curry last night and stew tonight. It would be good to stretch the legs, have a conversation wider than the tight circle of the three of us and provoke some informal contact with the locals. However, perception is all, and when we coincidentally passed, without any pre-warning, a short assorted column of BTR-60 wheeled APCs and assorted other low-value 'vege', an astute, alert and youthful Soviet officer leapt out, literally, in front of us, as the road conditions had forced us to a crawl, and we were 'detained'. It was lame, tame but unavoidable. But, again, from a purely personal perspective, a great experience.

Unable to swerve and avoid the situation without risking an accident and potential injury to the officer and other road users, we had to accept the fact we were quickly surrounded and blocked in, front and back – well and truly snookered. Tempers and levels of excitement were contained and professionally managed; we were not 'tarped', so we could enjoy the view, and no one attempted to wrench open doors or drag us out. Things very quickly took on the hue of the normal and the routine as though a detention on the streets of Halle was a daily and timetabled task for all players. The officer came to my window and I was able to make clear to him that

---

* Genscher is remembered in part for his disastrous coordination of the security force response to the Munich Olympics Black September attack, in particular for the carnage that occurred as the Israeli hostages arrived at the airfield. At that juncture in the history of counter-terrorist operations, institutional experience was underdeveloped, and the result fully justified that damning categorisation.

his duty was now to summon the local *Kommandant* and hand over control of the situation to him. I had already checked that all my tour observation gear was hidden, no map book visible, and Nick in the front seat had done the same. As we sat back to wait, we must have looked like the innocent liaison team that we were of course, unjustly prevented from attending to our mandated, authorised, accepted and age-old duties by the forces of communist aggression. A small good-natured crowd gathered and made themselves as equally comfortable to spectate as this episode of the 'West meets East' soap drama languidly unfolded.

In the event, there was less drama than a Honecker speech to the party faithful as the detention followed line for line and virtually word for word the exercise scenario I, Nick and our driver had worked through back on Salisbury Plain on our respective courses, back in the parallel universe of the free West.

The *Kommandant* duly arrived in his UAZ jeep. He was early 40s, dark-brown hair bordering on black with a Freddie Mercury-style moustache, which, unlike Freddie's, suited him well. He was an artillery major, as denoted by his crossed cannon 'collar dogs' with their black background and matching shoulder rank boards. He was smartly dressed in a standard light-green short-sleeved shirt, darker-shaded trousers and black shoes.

We commenced our conversation in broken German, with a smattering of English, through the crack above my slightly lowered window. No one in our G-Wagen spoke Russian. When it quickly became apparent that all was calm and kosher, I got out of the vehicle and we continued our chat under a beautiful cloudless blue sky, leaning on the G-Wagen's warm bonnet. He tried to make out that we had been taking undue notice of the Soviet vehicles, that we were not supposed to be here etc. etc. etc., all lines from the same script, but it was done without energy or conviction, and certainly without malice. And for his finale, he produced the *Akt* that I was required to sign, unless I reverted to my version of the script. This, of course, I did and we continued our polka for a little longer until the music stopped and we had both exhausted our parts, *Akt* signature-less. We chatted on a little longer with friendliness and openness about home and family, Halle et al. It was a good opportunity to air the old handling skills and attempt to build an initial layer of rapport, but of course with absolutely no expectation that this would influence the conduct of this absolutely standard detention, or lead to any future exploitation. Things would soon, however, change.

We received permission to move on; I provided some nonsense statement about our immediate intentions and destination and we parted company with a wave. No doubt for everyone, it provided some distraction from the monotony of life in the open prison of the DDR and the tedium of another 'training day' for the soldiery, and the crush did initiate a relatively rare phenomenon on an Eastern street: a traffic jam, the ubiquitous *Stau* of West Germany.

I consistently found our Russian hosts – as they were with our secondment to their in-country high command – to be polite and, once the initial caution and reticence lifted, warm and to have a good sense of humour. They were just as interested in us and enjoyed the novelty of rubbing shoulders with their sworn enemy as much as I did, and no doubt recounted the encounter to others with some gravitas and maybe a streak of personal pride. This brief episode certainly did nothing to change my positive opinion.

On the question of being detained, random and unprovoked though this example was, some tourers considered it a professional failure and a point of tactical shame. I certainly did not share that view. Anecdotally again, I had heard stories of some tourers of olden, golden days being detained some sixty times during a standard Mission posting. It was also true that some tourers who became quantifiable irritants to our hosts were aggressively targeted for detention to justify PNG-ing them and forcing their premature posting from the unit. This had happened more than a small number of times. But 'routine' chance detentions provided an excellent opportunity to converse with the enemy and, ultimately, demolish some of the preconceived barriers that existed, especially I think from their perspective. Later, any contact had the potential to develop in very different directions as we exploited the revised tilt of the Soviet orbit.

It was Mission tradition that a tourer's last tour, regardless of whether they were full-time tourers or part-timers, was punctuated by a farewell barbecue. The guest list was always invariably small and limited, not surprisingly, to the tour's three-man team – but, usually, less is more.

Wall down, with Gary's imminent posting out of the Mission, it was his turn. He organised for me and our mutually favourite driver to deploy with him on this special occasion – being the assistant ops officer, he had, after all, some sway in the matters affecting touring, and he had a very good working relationship with his boss, the ops officer, with whom, and his wife, I had incidentally been at university with. The day before we set out for the Bridge, we met for a quick shopping session in the US PX, down in their sector in Zehlendorf, to get some choice cuts of meat, some charcoal briquets and the other odds and ends we needed. This was the only acceptable place to go. A trip to the PX with all its American goodies, from meat to snacks to sporting goods, right up to Harleys, was reminiscent of a childhood trip to the toy shop. A great place and always a satisfying, and supremely value-packed, outing. We stowed the much-used portable BRIXMIS barbecue, probably also sourced at the PX, and headed off.

I don't remember much about the operational highlights of this tour, and needless to say we gathered nothing of any significant intelligence value, but we did have a bit of role and seat switching on day two. I had deployed in the back seat, with Gary acting as tour NCO. He swapped with me; Kev, the driver, moved sideways to the front passenger seat, and I moved forward, getting behind the wheel.

We had to navigate through several pretty tight, and bumpy, rough tracks, so I was able to put the G-Wagen through some of its paces. At one point we stopped to tackle a small driver training circuit. I don't remember the rationale for doing so, but Gary tasked us through the opening in the earth mound of the outer circle of the training 'ring'. Once in there, we were able to log the side numbers of a handful of BMP-2 APCs being put through their paces. As we prepared to exit, job done, we were seen by one of the APC commanders and recognised, not as a loitering UAZ, but for what we really were. The realisation seemed to upset him. A mad dash ensued for the gap in the Wall as the BMP driver 'floored' his accelerator. We all learned that afternoon just how rapidly a 'Bump' 2 could shift – they are fast. Behind the wheel, I met the challenge, and using our even more impressive acceleration and top speed, I timed it so we just pipped him to the post. We were out and away but had to endure a protracted chase as the BMP tried to pursue us.

That evening we stopped at the wooded spot we had decided upon during the tour planning phase, our tail well and truly clear of any surveillance, and the forest protecting us from casual observers. The meat sizzled and charred, and we may even have imbibed the smallest quantity of alcohol – I actually don't remember. We enjoyed a very tasty meal – the reality about eating outdoors is that even the most mediocre offering transforms into gourmet cuisine – and each other's company, all three of us blessed with robust, typically black, senses of humour. It was a fitting end to our momentous shared time in BRIXMIS – I will elaborate more on this theme in the next chapter. Sadly, we have since lost touch. I am pretty sure he moved to Australia with his wife and young daughter, on the back of an exchange posting there, and I hope life there is still good for him. I am going to track him down.

A common feature that affected all Mission touring, to a greater or lesser extent, was *Stasi* surveillance. Towards the end of the Mission's life, we were given an unprecedented insight into the wider modus operandi of the MfS and specifically their mobile and audio surveillance capability. I will cover this in more detail and have included one of my debriefs of 'Frank', a *Stasi* officer, at Annex A.

The first point to make on the subject is that, in tradecraft terms, their surveillance operatives did not employ covert surveillance against us in the sense that we certainly defined it. I very quickly confirmed in my own mind that they were what we referred to as 'narking'. What they did when they were behind a Mission vehicle bore none of the subtleties of covert passive surveillance that the Int Corps taught its operators who were trained to deploy against hostile intelligence service targets.* The *Stasi* teams did not care if they were spotted and hung on doggedly to their target regardless of their personal exposure. They were the proverbial dogs with the proverbial bone. So blatant and devoid of professional subtlety was their operational approach that I began to speculate whether

---

* The Int Corps ran what it called the Long Surveillance course, back at SIW in Ashford, equipping its personnel to maintain covert surveillance on the Soviet Military Mission operating in West Germany, and against other targets. This unit, known as 28, was not to be confused with the RMP team, the 'white mice'. The Americans had no equivalent of our surveillance team operating against the Soviet Mission in their zone, SMLM-F, until almost the last dying gasp. Nor did they request any help from the West German Federal Security Service, the BfV, in tracking Mission tours. In part, this counter-intelligence failure was founded on the complacent assessment, made and perpetuated by analysts in the US European Command HQ, USAREUR, that surely a small mission of only fourteen members could not pose any existential threat, compounded by analysis of its seemingly languid touring programme, and a desire at command level to protect USMLM in Potsdam from reciprocal action if the Soviets perceived that their own SMLM-F was being harried not until 1988 that this significant misappreciation was remedied by the creation of the twelve-man 'Det A'. Ironically, it was only fully operational in August 1989. By then it was only possible to claw back knowledge and understanding for a few months until the end of the Mission era, forty-two years after SMLM-F had become operational. I had attended the surveillance course only five years previously, in 1984, so was still current in terms of tactics and drills, both surveillance-wise and in terms of anti- and counter-surveillance.

in fact the 'real' surveillance was being performed by another organisation, covertly – perhaps a unit of the GRU, Soviet military intelligence, mirroring our own approach to SOXMIS with the overt RMP and the covert teams of 28. I had no proof to support this idea, just the conclusion that the DDR authorities were doing it so badly. In the event, I do not believe that any proof surfaced to support my hunch when the post-9 November flood gates opened, and formerly classified documents and information began haemorrhaging uncontrollably.

In tandem, bear in mind that car ownership in the DDR was way below the per capita norms of the west. Traffic density was light, especially in rural areas. This substantially aided identification of surveillance cars regardless of any other factor, but frankly on the ground it was relatively easy to spot them and any suspicion was usually very soon confirmed. Their cars, predominantly Wartburgs, Lada saloons and the excellent 4×4 Lada Niva, invariably contained three passengers – as 'Frank' outlined during our debriefs, one was the driver, the second the navigator and the third team member the radio operator. Operators were usually male – I never saw a female team member – and invariably sported leather or faux leather jackets and turtlenecks, giving them a sort of uniform Michael-Caine-in-*Alfie* look. Their mission was obviously to maintain tour vehicles under observation and catch us taking undue notice of military targets, ideally photographing a military target, a fixed installation or a vehicle or aircraft, and perhaps us photographing them. By 'narking' us, they were attempting to deter and prevent us from mounting intelligence collection, the proverbial itch that would not go away. Frank readily confirmed these conclusions. Generally their posture was unnaturally stiff and their apparent blasé indifference to tour vehicles was in marked contrast to the natural curiosity of someone on the street, who cranes their neck to observe something alien that does not belong in their realm. It certainly made them stand out. I am sure that some operators' skills were of a high standard but in the context of following Allied Mission vehicles, there was little scope for these talents to shine through. No doubt they were behind other targets that required being under foot surveillance, but against us we all remained vehicle-bound. It is probable, of course, that Mission members on cultural tours were followed on foot. This never happened to me, and I don't recall any reports of incidents that resulted if this was happening.

It has to be stated that not every tour found itself under surveillance, and because they, the *Stasi*, worked on a regional north/south basis rather than nationally, sections of a tour could find themselves being narked then, but on moving beyond the jurisdiction of the team, they might find themselves alone, tail-less. A national approach, as formerly adopted, according to 'Frank', would have provided the MfS with the ability to pick us up from the Bridge in Potsdam, getting off to a very good operational start, and maintain surveillance for as long as it suited their operational plan. But that was not how it was as we approached the final decade of the twentieth century.

Our brief was to tolerate their presence until we had specific target-related work to do and could not have them sitting watching us as we watched their military compatriots. Then we had to burn them off, wart-like, and leave them behind. To achieve this space, we had to enact fairly aggressive and obvious anti-surveillance drills, whether this involved high-speed driving to leave them behind – always potentially hazardous so not a drill to be embarked upon casually or lightly – route selection across tough terrain that would defeat standard saloons, violent changes of direction, U-turns etc. etc. Given the nature of the surveillance, there was absolutely no requirement for any skilful, subtle, covert drills – they knew that we knew that they were there.

I remember one day we were being followed but had some time before we needed to be on target, so spent it really aggressively shaking the team off that had glued itself to our bumper. I think that day I worked our way down through every drill on the list of 'How to evade an irritating MfS surveillance team', much to the driver's delight. It all looked very much like scenes from a bad Hollywood action film, but it was fun – there was probably no other theatre in the world where one could be as cavalier and overt in tradecraft terms.

What I will say, positively, about the *Stasi* teams was that once we had lost them, their 'lost contact drills' were efficient and effective. They would always eventually find us, usually relatively rapidly, and get back on our tails. I assessed this skill as being based partly on their obvious knowledge of where we might head to – they knew the targets that attracted us, had an analysis cell working in Potsdam to identify patterns and, of course, critically knew their ground as thoroughly as any team should – and partly on their genuinely effective and thorough drills, systematically searching for us after a loss. And, of course, not to be forgotten or underestimated, there was always the prospect of exploiting sightings of us being reported in by diligent, loyal party members.*

On a couple of occasions, on different tours, we had led our surveillants into enclosed wooded locations, where we intended to pause for a break. On one of these trips I got out of the vehicle and walked slowly and casually towards the 'trigger' vehicle that had followed us in, intending to attempt to open whatever dialogue I could – a pointless exercise. They were not entertaining any prospect of that and kept backing away out of range once I entered their 50m comfort zone.

On a rather more mundane theme, I had what we call an interview without coffee with the number three in the Mission, the (SO1 or senior ground tourer), one of only a few in my career, despite being a seasoned coffee drinker.

In common with all operational units, the Mission ran a roster of duty officers and duty NCOs as the first point of triage to respond to events that might, or demonstrably did, affect the unit. The duty officer role would be covered by all officers below the rank of lieutenant colonel, and there were three of these in the Mission including the deputy, who was RAF. The duty would run for a week, but on a daily basis a SNCO would be nominated to support the officer. In addition to the duty officer, there was always a Russian interpreter on permanent standby to support the Chief in fast-ball meetings with the Russian Command, and the standby tour crew – so on top of the busy operational cycle, there were always duties to eat into the limited amount of free time of some Mission personnel.

A classic example taxing duty officers anywhere would always be 'compassionate' situations relating to a serious incident, illness or death of a family member back in the UK, or elsewhere. Duty personnel would have the responsibility of passing on the distressing news to unit members and

* Had 'Frank's' debrief come earlier in the Mission's history, it would have served to better shape our approach to the MfS, and the GRU/KGB, and been the catalyst for a more systematic and formalised defensive strategy. But ultimately it did validate the Mission's cautious tactical approach, both out on the road and in the House, in terms of the measures that we did take across the spectrum of security risk. In reality, as it turned out, we had got things pretty much right where it really mattered.

then keying the necessary procedures to arrange transport for a rapid movement home, if required.

Back to my coffee-less interview: I was on duty with my deputy, a senior and extremely experienced and competent Int Corps WO1, who was the 2IC in the Weapons office. The message had come from the Mission House that a tour vehicle had broken down in the local area, so just down the road from the Bridge. Obviously the tour had been able to use a local DDR landline to pass on news of their issue to the House. A clearly established SOP was now triggered to transport a standby replacement tour vehicle, probably an Opel Senator, across the Bridge, RV with the stranded tour crew, and make the exchange, thus allowing the tour to continue on their way whilst recovering the unserviceable G-Wagen on the Mission-towed trailer back to the workshops in Berlin. I was familiar with the SOP, as I had been on the receiving end of it when my vehicle had also suddenly and unexpectedly broken down – I have to say this was not a common occurrence. Luckily, this particular rescue mission did not have to drive miles and miles down to the Czech border or some other far-flung region of the DDR.

I spoke with my deputy and ascertained that he was happy and confident with the situation and that there were no complicating factors that had to be vectored into the decision-making process – for example, the aftermath of an accident or an injury to deal with. Reassured that all was under control, I requested that he inform me once things had been resolved and the matter had been successfully brought to the conclusion that we both anticipated. And so it came to pass.

The next morning I was summoned to the SO1's office and given my interview whilst standing, at least no coffee cup to balance, and told that I had failed to take my responsibility seriously enough etc, etc. In some situations it is best to trust the book inserted down the back of one's trousers. This was such an occasion. However, inwardly I was fuming at being treated like some operational simpleton, but it was just not worth a comment that would of course have created a situation where one really did not exist, and where things might have escalated to silly altitudes with even sillier consequences. I have since told a boss in civilian life that he was simply not a leader but this was just not the thing to do in the military. It was, however, an interesting exhibition and insight into the senior Army ground tourer's appreciation of his men. He appeared to believe that a senior warrant officer class 1 in the Int Corps was incapable of following a well-established and codified administrative procedure. As an infanteer, this was akin to him doubting that the regimental sergeant major (RSM), the senior soldier in his battalion, was capable of wielding similar responsibility – this would have alienated the entire warrant officers, as well as sergeants' mess with a resultant, and reciprocal, backlash. I do not know whether this petty snapshot signified dislike for me, the WO1 or our shared cap badge. But I did witness another example of what I will call petty short-sightedness when he refused to endorse an application for commissioning from another senior Int Corps warrant officer – he accompanied it with a snide remark as to whether I could see the individual sitting in the officers' mess. Frankly I could and had seen many commissioned SNCOs in the Corps who were less professional than this first-class soldier. So simple an exercise of prejudice effectively ended this senior professional's career as he approached the twenty-two-year point that signalled enforced retirement. It was a great shame and not a fitting or appropriate way to recognise the service and commitment of one of our military colleagues.*

One treat I vividly recall was going up in the Mission's Chipmunk for a flight. I have already outlined the background to OP OBERON. It was an invaluable asset within the 20 statute miles of the Berlin air safety centre, providing a unique perspective on activity in the local area. Photos taken over-wing provided a full and unobstructed view of targets on the ground and, from the technical perspective, generated a completely different profile of target equipment, capturing overhead shots otherwise only obtainable from chance shots from bridges or overpasses. On this flight, however, we did not deploy to gather intelligence as our primary aim.

Tony was the squadron leader 'special aircrew' pilot who commanded the OBERON flights and, as denoted by his status and seniority, he must

* Senior non-commissioned officers had few options to serve on beyond a maximum of twenty-two years. Officers, then, were forced to retire at 55 unless they made general. In the end, I retired early at 45, ten years prematurely, frankly tired and in need of a change.

have ranked as one of the RAF's most experienced aviators. He was also, typically, a super guy.*

We met up at RAF Gatow, strapped on our parachutes and off we went. We lined up on the main runway as Tony went through his brief pre-flight checklist. With the throttle open wide, we were airborne in no time, with less sense of acceleration and movement than on a Vespa. I had already had a bit of stick and rudder time on a Bulldog – and would later perform a few take-offs and landings in a small Cessna – so I had some idea of what I was doing when Tony passed the controls to me.

Our bird's-eye view looking down into the local area's barracks was very special: row after row of parked up vehicles and ant-like Sovs shading their eyes, looking up at us, no doubt fully aware of who we were. Our unique perspective reinforced the peculiar nature of many Soviet barracks and training areas. They were well protected at the front with formal gates and wire, guard posts and all the other trappings of overt protective security measures. The rear, however, was often wide open and thus it was potentially very straightforward to enter sensitive areas without realising that one had done so. This became important to me on a couple of occasions as the new year unfolded.

In the past, there had been numerous incidents when irate sentries had fired at the plane, occasionally punching holes in the fuselage, but no Nicholson-like fatalities, thankfully. We were unmolested. The highlight of this 'sortie', however, was performing aerobatics over the DDR, looping, rolling, diving, flying inverted, the whole playlist – I have to admit, with Tony at the controls, not me – before we returned to base. It is the unarticulated aim of pilots in these situations to induce fear or at least nausea in their earth-bound comrades, but for me it was a great and novel experience to be cherished, and a great morning out.

---

* Spec aircrew were those pilots who made a career decision to drop out of the standard RAF career rat race whereby flying jobs were interspersed with staff appointments to prepare officers for senior leadership positions. Instead, the deal was that they spent the rest of their careers in the cockpit in return for an enhanced salary hike but the prospect of stagnating rank-wise as a squadron leader, a major in the Army. I thought it was a good pragmatic scheme and maximised the value of what were, in reality, scarce and valuable assets. All the other pilots in the Mission were career fast jet pilots, with the exception of one C-130 navigator, very much the odd man out. Had I been in Tony's position, I am sure I would have done the same and followed the same career path. You don't join the Royal Air Force to fly a desk – or, as was offered to me when I passed RAF officer selection at Biggin Hill (as an insurance policy if things had not worked out with the Army), to join the RAF Police or RAF Regiment.

11

The Beginning of the End

The week beginning 6 November 1989 kicked off, for us, like any other week in Berlin.

Summer was over and the prospect of a cold, crisp winter beckoned as warm jackets were removed from cellars, given a shake and an airing, replacing lighter wear. Otherwise it looked like business as usual in the run-up to the lights, trees, Christmas markets and unique festive atmosphere of another traditional German Yuletide in the former capital of the Third Reich. Maybe, I'm sure we speculated, there would be the prospect of some snow to complete the idyll.

I was back on pass and was scheduled to begin an extended local tour with Gary. We had prepped as normal. The latest local tours had reported nothing new; all appeared normal and routine in and around Potsdam and its local garrisons.

Of course, as the raft of contemporary documentaries analysing this time remind us, the wider DDR appeared less settled. It was undoubtedly, so it seemed to us, experiencing one of those periodic episodes of increased volatility that kept reappearing to test and solidify the Stalinist credentials of the party leadership. Erich Honecker, the country's leader, had begun the year stating boldly in a speech that the Wall would still be standing 50 or even 100 years in the future – he clearly had absolutely no conception of how rudely he had prodded and tempted the fates. But a groundswell of anti-regime sentiment had begun to slowly but steadily assert itself. Honecker's finger was firmly and reassuringly maintaining pressure in the DDR's dyke, but fledgling cracks had begun, in parallel, to be exposed in those of neighbouring less resolute and less staunch pact members, crucially in Hungary. Here, spring had been marked by the dismantling of the border wire with neighbour Austria, perhaps whilst engaged reminiscing about the faded glories of past Hapsburg solidarity. Whatever the motivation, increasing numbers of DDR citizens had begun making their way to Hungary, exploiting their right to travel within the Eastern Bloc, and had slipped unobtrusively across the now de facto open frontier into south-eastern Austria. Coinciding with the accession of Poland's first non-communist prime minister since 1945, weekly protests began to gather momentum in Leipzig, pushing for the modest goals of freedom of travel and free assembly. Another hole slowly but resolutely started to widen, threatening to challenge the integrity of Honecker's dyke. It too was subsequently quickly plugged, but not before East Germans were rushing to Prague to take refuge in the Federal Republic's embassy there. The nightly news programmes showed film of East Germans struggling to find space in the massively over-crowded embassy grounds, giving courage to others to push their luck.

The ripples of change swelled, transforming into veritable waves with Gorbachev's visit to East Berlin, in October, and his unrehearsed call for reform. His words resonated more symbolically, coinciding as they did, with this sacred month in the communist calendar. The SED regime pushed back with the righteousness of zealotry, reversing its velvet glove policy of toleration regarding the protests in Leipzig and revealing the iron that had always underpinned the actions of the state. The prospect of violence appeared to be a realistic scenario as the protests swelled and assumed an apparent life of their own and, in response, the numbers of security force personnel cordoning the protesters burgeoned. It appeared the city was one spark away from a real and deadly conflagration.

But ironically, and rather shockingly, the first fatality was Honecker himself as he was forced into shocked retirement, eleven days after Gorby had returned to Moscow. He was replaced by the unsavoury-looking Egon Krenz. I always felt his features bore some Hammer House of Horror-like resemblance rather than the reassuring countenance of a politician ready to lead his people away from the brink of potential disaster and into an era of as yet uncharted reform.

On the Saturday before we left for Potsdam on our tour, 500,000 gathered in East Berlin's Alexanderplatz to protest in favour of enhanced democracy. Things were looking yet more serious and the knife edge upon which affairs balanced appeared to be progressively and irretrievably thinning. We, in the Mission, of course, noted this. We were cautious but not overly concerned at the size of the gathering or the trend that appeared to be unfolding and gaining momentum. We reminded ourselves again that this was not the first time in the regime's history that there had been in-flight turbulence and periods of instability and uncertainty. In the longer run, we recognised that these aberrations had always passed and level flight resumed. We expected the status quo, albeit an essentially cosmetically modified one, to reassert itself this time too. Certainly, we assessed that Potsdam would not be affected by any likelihood of mass demonstration and change. No doubt the same conclusions were being drawn by our French and US partners, and our Soviet colleagues, in and out of the intelligence community.

We passed over the Glienicke routinely, called in at the House and got the tour under way. The mood was good, the team clicked, and the prospect of a couple of busy days and evenings interspersed with a good bed and breakfast made for high morale. We were still some months away from Gary's farewell barbeque, so he remained buoyed in pursuit of that final scoop.

We worked through our list of targets. All continued as dictated by the precedents of operational routine until, superficially at least, we picked up a story on the radio news that appeared to reinforce the trend that political unease in the DDR was continuing to rock the SED boat. The government was reported to have resigned. We paused, pondered but ultimately thought little of it and, as we drove our routes, our view on the subject was seemingly confirmed by the normality reflected in the everyday activity of the local people. Scenes of routine being maintained greeted us wherever we looked, on the streets, outside the shops and everywhere else we turned. Predictably, the Soviet military machine remained completely insulated from any repercussions of internal DDR political activity, and they continued scrolling down through their lists of routine things to do, just as we were.

The next day, our equanimity was ruffled a little more vigorously with a second news story that another more serious resignation had occurred in East Berlin. Now the cabinet had resigned. Surely this was a more noteworthy signal of some dramatic political undercurrents that Krenz's assumption of power was not yet successfully stabilising. But again, if indeed the DDR news was reporting the story, it had not yet provoked little huddles of excited comrades to gather in a highly charged, speculative mood, or indeed any other mood. Normality remained the order of the day. We were slightly puzzled but mirrored the mood music and continued touring. We sat and watched a small column of 2S1 122mm self-propelled guns pass us and photographed and recorded their details – no need for questions; they were obviously local. We pulled out of the cutting on the roadside we had slipped into and figuratively marched on.

The tour concluded on time, 9 November, without incident or note and we re-passed across the Bridge mid-afternoon after a late lunch with absolutely no sense of what would be happening less than six hours hence. We paid our courtesy call at USMLM, left our brief tour highlights and returned to the Stadium and, after de-kitting and handing in film, home.

I got a call from Gary just after 7 p.m. I had just eaten and was continuing to sort out my kit, getting ready for the next planned tour early the following week. As I listened on my end of the line, he was not his normal laid-back, calm self. His speech was rushed, his manner excited.

'Have you heard the news?'

'What news?'

'The Wall.'

'Uh?'

'The Wall is down!'

'Pardon, what wall?'

'*The* Wall!'

Those brief, short, monosyllabic words recorded and broadcasted the end of the old world that had fed us at lunchtime and marked the start of the bright new one that would provide our breakfasts, lunches and dinners for a very long time to come.

It was difficult to take in and process – the short words were too big, too imprecise, too, frankly, unbelievable. Gary confirmed that, incredibly, yes, he was referring to *the* Wall, the Berlin Wall. But clearly the Wall had not physically collapsed. That would have required an awful lot of plastic explosive and engineer plant and was not the activity of a Thursday autumnal evening. There must be some other explanation. Gary continued his report, now regaining more of his usual composure: crowds of East German citizens had started to build up on the Potsdam side of the Glienicke Bridge and a steadily growing trickle was reported to be coalescing into a wave that was seeking to push through the Russian military security on the Bridge. There were similar mass gatherings at the main crossing points through the Wall on the eastern side of Berlin city centre. He suggested we get ourselves down to Glienicke to assess and report on the situation and see what on earth was happening.

For a brief instant I was paralysed as I stood, disconnected phone still in hand, processing Gary's blockbuster recital, trying to order the flood of implications that swirled shapelessly and formlessly around in my head. It was not 1 April and anyway well and truly after 12 p.m. Gary was a cautious professional and was clearly forwarding news he had been briefed on by dint of his role in Ops. This was a serious office that did not trade in maybes, rumour or speculation. Something monstrous was clearly happening in Potsdam, and elsewhere. Something inconceivable had clearly happened *by* the Wall, *at* the Wall, *to* the Wall. We had to be there.

I grabbed my jacket and a small pocket camera and raced around to Gary's married quarter to pick him up. We made the Bridge in record time, parked up in the parking area set aside for the relatively small numbers of tourists who made the effort to visit the area to stare, paradoxically, westwards across the water into this unspectacular corner of Potsdam, and the immovable DDR. On our side of the Bridge, all was quiet, but the sight that met our gaze, across the darkness, was truly unbelievable and beyond rational comprehension.

We did not then know the sequence of events or the timetable that had created this backdrop, but the surreal activity we surveyed was very, very real. We numbly took in these images that would become indelibly chiselled into our consciousnesses as the next few hours unfolded, and will stay with us both for the rest of our lives. Not for the last time, I distractedly removed my camera from my pocket and took some pictures. I clicked back into professional mode now and began to wish I had brought one of my SLRs from home to maximise the quality of these clearly momentous images that I was recording.

We had reacted with the speed of an ambulance crew called to a motorway pile-up in response to the press conference, held between 6.53 p.m. and 7.01 p.m. in the DDR's International Press Centre in East Berlin, at which Günter Schabowski, instrumental in the overthrow of Honecker, had hurriedly read a pre-prepared press statement, handed to him by Egon

Krenz just as he went live, detailing revised travel regulations. In response to probing questions from one of the foreign journalists present, Schabowski had rapidly re-read his script, shrugged and, after a moment's hesitation, announced clearly but unbelievably that, with immediate effect, all crossing points into West Germany and West Berlin were open. The Wall started to wobble.

The events we were to witness this coming evening and over the next few days ensured that it crumbled and toppled, lying broken, metaphorically, on the floor, completely beyond any prospect of restoration and repair. The Wall had been breached and was beyond resuscitation, now useless and surplus to requirement. Once breached, it could not be re-sealed. It seemed that, incredibly, Gary and I had been the last Allied officers to re-cross into the west from the 'old' DDR, a country, and a political situation, that would now be changing faster and faster as each new day successively dawned on the last, like a dervish spinning out of control until, like its Wall, it too fell.

The bridge we surveyed in awe was not our Glienicke Bridge. Normally ghostlike and empty, illuminated by hazy, smoky spot lamps, it was now a writhing living creature as a mass of laughing, crying, shouting Potsdamers jostled, good-naturedly and excitedly, the bewildered guard force – Red Army and East German *VoPo* police – impotently trying to hold them back whilst completely unsure what their orders were in this new, snap-imposed, seemingly headless regime. Were they required to restrain the crowds, disperse them, or were they authorised to open the crossing point in line with the host government's latest proclamation, processing paperwork or allow themselves to be swept inconsequentially aside? We should have sympathised with their predicament but we did not. We watched on.

Beyond the press of humanity, a line of cars, predominantly *Trabis*, waited patiently in line to exploit Schabowski's shrug. They had waited twenty-eight years for this moment. What were mere minutes more? We could see DDR flags being waved and hoisted aloft. The crowd was definitely jubilant, one huge living, beaming smile, supremely good tempered in their elation. And it was elation, boundless, brilliant in its intensity and honesty. I knew from experience that the collective mood of crowds is fickle and can turn to murder on the edge of a coin, but this crowd was fuelled by the most benign emotion experienced by man.

Gary and I approached our end of the Bridge and started to move across it – approaching the bold words of the DDR border sign, '*Deutsche Demokratische Republik Potsdam*', and the hammer-and-compass-wreathed symbol displayed between country and city names – and onto the Bridge proper where, just earlier that day, we had sat whilst the Soviet guard disinterestedly, another day of national service soon to be over, booked us out of the country and back into the west. It was bewildering, emotional, stunning, amazing, surreal, unearthly. My head filled with an endless stream of words, colours, adjectives. It was all unquantifiable and indefinable. When I have tried to describe these events to family and friends over the years I have found it consistently difficult to avoid talking in clichés but the sheer intensity of emotion was something I had never experienced before or since. The laughing, crying, shouting was an endless, rolling cacophony of sound but somehow mellow and musical with none of the raucous, hard edge that characterises the shouts of a football crowd, a political demonstration or the baying of a riotous horde.

The cordon of *VoPos* and border guards finally fractured as their indecision left them airless and directionless, and the floodwaters swept forward across the final span of the bridge and into free West Berlin. We stood to one side, taking in the beginning of the end of state communism, not just in this divided Germany, but in Europe as a whole; to misquote and disagree with Churchill, it was certainly not the end of the beginning.

Change, change and more change was to accelerate hard on the heels of this evening. It was to affect Gary's life, my life and the lives of the rest of the Mission members. In thirteen months the Robertson-Malinin Agreement would be history, dead, lifeless words, and so would BRIXMIS and its US and French counterparts – and our Soviet comrades in the West. Most of its members would be equally gone – only a hard core of us would remain, reconfigured, retasked and re-energised.

But that was in the future as the crowd surged forward, savouring their freedom, now travelling freely for the first time in the country's history, heading towards Berlin's city centre on buses and taxis that had now began to congregate to transport them, free of charge, off to see, hear, smell and feel the sights. The line of cars began to crawl onwards. The genie was out of the bottle and gone. Where only G-Wagens had formerly had right of way, this armada of primitive, sub-standard machinery, horns tooting, flags waving, crossed and began their migration towards the Ku'damm, the zoo and all the other landmarks that their occupants had only seen on West German TV, before this moment of ecstasy. Absolutely breathtakingly amazing.

I don't remember how long we remained drinking in this intoxicating cocktail of sound, sight and emotion, but eventually we tore ourselves away and followed the masses into the *Stadtmitte*. It was the most energising, throbbing, vibrant party I have ever witnessed – the most intense celebration of joy and just sheer, pure enjoyment of living. Having been trapped in *Stalag* DDR and denied the opportunity to travel westwards, these people now had hope for the future and were able to express and articulate that hope. Had Honecker addressed the simple issue of controlled travel beyond the boundaries of the socialist world, he might still have had the reins of power between his fingers, and this 9 November might have been a quiet, ordinary autumn evening rather than the first day of the new tomorrow.

I don't remember when we got home to bed but it was not before we had joined the flock to the Wall where it swept gracefully around the façade of the Brandenburg Gate. Those crowds coming eastwards from Potsdam and beyond merged with those who had breached the Wall from the east of the city. I wondered just how many extra temporary citizens the city hosted that evening and on the ensuing evenings and long nights as the situation began to normalise. Numbers beyond casual estimation for sure.

Next morning in the office, the shock of events was still as raw and the novelty just as new. To say that the system was reeling was a lovely piece of succinct British understatement. Conventional wisdom was in tatters just as much as the confidence of the East German hierarchy. No one, *no one*, east, west, north or south, seriously believed that change as radical and unquestionably far-reaching would have occurred during our tenures in the Mission, the Army and probably during our lifetimes. How the leadership in the east was processing events could only be speculated at. I am sure our leadership was in constant touch with the top generals back in the zone in West Germany and with the FCO in London, just as the lines burned hot to Washington DC, Paris and no doubt Bonn and, maintaining the true spirit of reciprocity, in the other direction to Moscow. Discussions about the future must have already begun as contingencies and possible scenarios were exposed and debated – this was the way of the military and the way of government. These discussions will have been most focussed in Bonn as Chancellor Kohl locked onto the scent of immortality that would shroud the man who united Germany. But it was clear this was unlikely to have been a scenario that would have been table-topped recently in any of the relevant forums. A rare opportunity to express new fresh opinions and ideas had confronted the west's leadership from nowhere. It was clear, however, that the prospect of the SED government regaining control and re-imposing the old order was not a realistic scenario. The bull had tossed its cowboy and had escaped the pen and got clean away – no one was going to stop it and tame it soon.

At the lowest tactical level, I was briefed by the SO1, my friend, the senior Army officer in the Mission, to get myself back downtown with an F3 and start snapping to record events as they continued to unfold along, and very soon on top of, the Wall.

I booked out a camera and medium-zoom lens, stuffed my pockets full of colour film and headed off on the *U-Bahn* to join the rapidly increasing press corps presence that had begun to descend vulture-like on the city as they do in times like these – a story to flog until even the bones of the horse could take no more.

I took in the sights, sucked up the atmosphere and took an awful lot of pictures. I thought it was a fairly straightforward and effective transition from intelligence officer to photo journalist. I have often made the point that journalists and intelligence officers are both engaged in the same business, just for different bosses, to achieve different ends and on different pay scales. I was certainly not alone as the area around the historic and iconic Brandenburg Gate was jammed with news vans, sat dishes and other reporters with their Siamese twin camera men. But the stars of the show were the thousands upon thousands of East Germans still flocking through the breached Wall, many now only identifiable by the strikingly odd appearance of the denim of their jeans or jackets, mixing freely and harmoniously with their western cousins and the flocks of other birds who had descended upon the city to taste the making of history. I wondered what stories the news reports on state television the other side of the Gate were airing. I could not believe they were ignoring the current blockbuster news, or were able to spin it in any other way than admit to reality.

I returned that evening and the next day and night. It was more of the same except that now the Wall had been made more accessible, not least its very top. I too joined the revellers to look down from atop it. Surveying the sights from this vantage point that no one would ever have believed possible, I could not for a long moment process the significance of where I was. I was *on* the Berlin Wall. I had thought it revolutionary enough to have pried off my chunk of raw concrete ten years before. I never imagined any points of light converging to produce this picture. In fact, I don't actually remember ever considering the prospect of the Wall ceasing to perform its role. Perhaps my view was limited spatially but what it represented was truly momentous. We could have been at the very summit of any unclimbed Himalayan or Andean peak. The panorama that spread before us had never been witnessed by man. Looking down, we were recognising the past; levelling our gaze, we were unequivocally glimpsing the future. It was incredible to think that a week ago we would have been shot dead for what we were now freely and irreverently doing, totally unmolested by eastern security forces. How did the ghosts of those who had lost their lives in their tragic, forlorn, failed escape attempts regard this freedom?

People jostled and pushed to create space both at the foot of the Wall, as they prepared to scramble upwards, and once up there, but it was without malice or frustration. Again, patience seemed to steer, but not limit, emotion. On the Wall, there was continual movement as the mass adjusted minutely to the limited space – it perversely somehow reminded me of the tiny movements emperor penguins make as they tap their feet as they warm their eggs. People atop the Wall reached down to haul up new summiteers whilst others below pushed and lent shoulders and hands. Any other situation would quickly have descended into impatience, tension, outburst and likely some limited violence, but this wave of humanity maintained its almost euphoric state of peace, goodwill and grace. We had proved that there was the reality of 'people power' and this has remained as a new force in the international system to this day. Were we also proving that mankind is capable of co-existing in peace and harmony? Yes, undoubtedly, but perhaps for a fleeting week, maybe a month, until the novel or abnormal morphs into normal routine and the return of cynicism, greed, self-interest and competition triumphs.

More sonorous and melodic dialectal notes now mixed with the harsher tones of north Germany. They in turn were already being joined in regular counterpoint by the incessant woodpecker-like tapping as others took hammers and chisels to break off pieces of the concrete dinosaur, ensuring it was dead beyond doubt. This incredible spectacle was continuing to draw people from far and wide, eager to share this moment of unrepeatable history. I have often wondered how many romantic unions and dalliances were these days inspired as they unfolded and slowly set their own routine. I took one last look around, eastwards and back over my shoulder towards the lights that somehow now were not quite so bright or beacon-like. Now it did not matter which side I eased myself down onto, but I dropped westwards and, pushing through the seeming unending tide of newcomers, continued to breathe in the heady air of our new Berlin, heading for home.

A bird's-eye view of the Berlin HQ building. The Mission occupied the top floor.

An aerial view of the Glienicke Bridge, 'The Bridge of Spies', looking into the East.

The rear façade of the Potsdam Mission House from the lakeside gardens.

A BRIXMIS G-Wagen posing unobtrusively in the trees.

A G-Wagen from the rear, the Mission plate almost completely obscured by mud.

The author with a trio of fraternal comrades. Visible in the front passenger seat is one of the duo of tourers who was later struck down by possible chemical weapon poisoning.

A BRIXMIS Opel Senator.

A Soviet Pass, tour uniform arm badge and No. 2 uniform 'key fob' from over the left breast pocket.

The author atop a G-Wagen to get a quick shot over a perimeter wall.

Only a BMP-1, but a graphic illustration of just how close a tour could get to the enemy.

MT-LBu on rail flats.

A rail move of T-64 MBTs.

A lone 2S3 SP artillery piece. The seeming mod above the turret top is not a scoop but rather a railway siding light on the far side.

The OP OBERON de Havilland Chipmunk, in the hands of a BRIXMIS pilot as capable of a recce platform as a Tornado GR4A.

A Stasi operator or nosey 'Eastie'.

On the Bridge, the Soviet guardroom to the left.

SOVIET EXTERNAL RELATIONS BUREAU

(As known to BRIXMIS - Jun 90)

ZOSSEN OFFICE

Chief of Bureau

KUKLIN
Vladimir Sergeievich
Col - Arty on Black
Married accompanied

KUKLINA
Valentina Vladimirovna

UI Staff Officer

VOROB'EV
Vitalij Nikolaevich
Lt Col - MR on Red
Married accompanied

VOROB'EVA
Galina Fyodorovna

Senior Interpreter

SAVCHENKO
Sergej Danilovich
Maj - MR on Red
Married accompanied

SAVCHENKO
Lyubov Mikhailovna

POTSDAM OFFICE

Deputy Chief of Bureau

POLOZOV
Robert Ivanoich
Col - MR on Red
Married accompanied

POLOZOVA
Tamara

DEMCHENKO
Vladimir Georgevich
Lt Col - Engr on Black
Married accompanied

DEMCHENKO

OSTROUMOV
Sergej Nikolaevich
Maj - Arty on Black
Married accompanied

OSTROUMOVA
Nina Sergejovna

Interpreter

STYOPKIN
Mikhail Ivanovich
Capt - Arty on Black
Married accompanied

STYOPKINA
Marina

Interpreter

KROMORENKO
Eduard
Capt - MR on Red
Married accompanied

KROMORENKO
Tatyana

Finance Officer / QM

ZIKOV
Sergei Ivanovich

ZIKOVA
Tatyana

Other Frequent Contacts

Commandant Checkpoints

LYUDOMIRSKIJ
Yurij Anatol'evich
Col - MR on Red
Married accompanied

LYUDOMIRSKAYA
Tamara Mikhailovna

UI Staff Officer

PLIEV
Yurij Mikhailovich
Col - Arty on Black
Married accompanied

PLIEVA
Lyudmilla

Commandant Potsdam

SHEVCHENKO
Viktor Ivanovich
Col - Signal on Black
Married accompanied

SHEVCHENKO
Valentina

Members of the Soviet External Relations Bureau (SERB) in the summer of 1990.

The author at a social function in Potsdam with Sergei Ostroumov (right), Yuri Pliev (left) and Robert Polozov (background), Deputy Head of SERB. All three were likely GRU officers although it was reported that one member of the Bureau (not identified) was a serving KGB officer.

The author with KGB Major 'Steve' Maltsev having spotted him amongst the crowd at a Soviet air base open day. His state of shock and surprise was palpable!

The tidal wave of euphoric Potsdamers crossing into the West on 9 November as the Wall ruptured uncontrollably.

More the following day.

A BRIXMIS cultural tour laying a wreath at the desolate windswept Buchenwald Concentration Camp.

12

A Brief History of Time

BRIXMIS was unique as a small intelligence-gathering unit that remained essentially unchanged in terms of organisation and in terms of tasks for forty-four years. In size, at its largest, the whole Mission, with its disproportionately large administrative tail, did not exceed 100 servicemen and women. That is smaller than the four platoons of an infantry company.

The Mission punched way above its weight and it landed some amazing blows and collected some really valuable and weighty trophies. Above everything, the ability to watch and then evaluate the quality and effectiveness of Soviet military training provided unrivalled insights into the real menace that the Bear represented. Tourers were able to see through the propaganda and the politicisation that regulated the Cold War and evaluated the Red Army capability for what it was, not how it was presented. A conclusion as valuable as it was significant, was that the Group of Soviet Forces stationed in the DDR was massively under-resourced in terms of its manpower establishment – some units were manned at levels as low as 50 per cent of their officially designated strength. This was arguably intelligence of critical value and debatably unobtainable from any other source including HUMINT, which likely would not have been believed.

Tours collected intelligence on new weapons systems and platforms, validated tactics and doctrine and confirmed numerical estimates on equipment holdings and the organisational formation of units. But above that, they provided commentary that informed assessment as to how effective the Warsaw Pact would be if it punched west and struck at NATO's defensive positions. These assessments guided the judgement that the Warsaw Pact certainly posed a less than real conventional threat. This was, however, significantly tempered by the reality of the battlefield tactical nuclear weapons and longer-range ballistic missiles that supported ground forces. And BRIXMIS was able to confirm their reality.

In this way, the technical effectiveness of equipment, the doctrine governing its deployment, the utilitarian approach to function and the sheer scale of resources deployed were confirmed but critically tempered by qualitative analysis. Strengths were offset and counter-balanced by the weaknesses that were apparent to the teams who stood off clandestinely observing training. The Soviet lack of doctrinal and tactical flexibility was apparent, the centralised nature of command and control and the absence of 'command leadership' and initiative at lower levels of command were clear to see. These 'footnotes' were just as important as samples of new-generation tank armour or pictures of the latest surface-to-air missile system – in the event of conflict, arguably even more important as shapers of defensive strategy.

The modern Russian Army is not the Red Army of Cold War days but military ethea, cultures and histories exist, mark and shape not just the past but survive into the present and extend into the future. Today's units in Ukraine have inherited not just formation designators – names and numbers – but certain ways of thinking and acting. With the huge reliance on conscripts – poorly motivated, poorly trained, poorly disciplined, poorly resupplied and, arguably, poorly led – almost obliterating the professional face of the armed forces post-USSR collapse, a drone's-eye view of fighting forces would more persuade the observer that he is looking back in time to pre-1989 days.

It is these weaknesses above that we appear to be seeing in the 'new Red Army' that is currently struggling to achieve its operational goals in Ukraine – the downside of cultural continuity. It is still beholden to its Cold War heritage and approach to battle, not least in terms of the poor standard of soldiery and the lack of flexibility once the sharp operational edge is blunted and disrupted. The number of Russian generals killed in forward positions, at the time of writing, attests to the re-emergence of the doubts we in the Missions had about effective command and control. It appears that the lessons shaped on the training areas of the old DDR can still carry water.

I find it fascinating to reflect on how the Mission evolved over this period from those early hesitant first steps, prospering as the Cold War cooled and tension between the two blocs hotted up. It is equally fascinating to see how its two roles self-adjusted, evolving without deliberate direction and then flipping, one replacing the other as the key function as the wartime relationships themselves evolved, liaison very quickly relegated to a secondary but, it must noted, certainly not irrelevant role. It is fascinating to witness how tactics and operational procedures and processes developed and matured. And, above all, examining the broader picture, it is intriguing to record and evaluate BRIXMIS's contribution to the enduring peace that has outlasted the Cold War, and the role it played at times of enhanced tension when the smallest miscalculation, blind spot or miscommunication could have sparked the third, and almost certainly final, global conflict. Then there were the scoops, those exciting sightings of newly deployed equipment, either the result of chance encounters or carefully targeted collection, and the steady flow of technically based intelligence that drip-fed the Allied scientific community.

Throughout the Mission's life, after the immediate post-Second World War decade had passed, there was a relatively regular and even rhythm of scoops. It is interesting to remind ourselves that, when we look back retrospectively – with a cavalier tinge of dismissiveness, especially towards those formative days – although we are surveying veritable antiques, like the AK-47, the JS-2 heavy tank and the T-54, at the time, first sightings were coups that were greeted with excitement and approbation for the 'scoopers'. These were a new generation of war fighting tools – bright, shiny and deadly – assessed as being just as important as the introduction of a modern stealth fighter or a generation of drone today.

It did take some time before a coherent and cohesive set of SOPs emerged. But their codification gathered pace and momentum once there was institutional clarity that the Mission had to play its role in the gathering of intelligence against the Soviets to supplement the collection efforts of the strategic agencies. It was this fundamental dovetailing into the mosaic of high-level intelligence collection that marked out the Missions as the special units that they were. In terms of what the tours brought back to Berlin, BRIXMIS was the golden goose that was able to plug gaps in the various staffs' and MOD intelligence directorates' collection plans that no other unit or agency, with operating budgets way in excess of the Mission's, could fulfil with sustained and systematic regularity.

Early tours recognised the importance of rail transport and the strategic-level intelligence that monitoring important lines could generate. This was, in reality, nothing new as the teams from SOE and other clandestine units had engaged in similar activity during the Second World War. But interestingly, BRIXMIS's early rail watches focussed on economic intelligence collection, rather than seeking to support the effort to subordinate Soviet units, recording, for example, steel deliveries entering the eastern zone from the USSR to augment the reduced, and reducing, output of the east's steel mills, due in large part to the Allies' reciprocal blockade, post-airlift.

As the Berlin Blockade had threatened to throttle the city into submission and reset the new post-war balance of power, the Mission was not idle. The massive armada of flights generated by the Allied air forces inevitably resulted in regular air accidents and a considerable number of fatal air crashes. It was the Mission's rather morbid task to locate the crash sites and participate in the task of identifying and repatriating dead air crew. Between July 1948 and the following August, there were over forty accidents involving just transport aircraft of the RAF alone.

With the air bridge successfully maintained and ended after the Soviets backed down and ended their blockade, the Mission's interest in Allied aircraft did not cease. It continued into the 1950s as a number of aircraft either crashed accidentally or were shot down, having strayed inadvertently into eastern airspace, or whilst engaged in deliberate reconnaissance overflights.

By the time of Queen Elizabeth's coronation, a clearly identifiable touring pattern had emerged and was guiding BRIXMIS teams' intelligence collection. This was based on a clear appreciation of the nature of targets that revealed the type of intelligence that the Mission could collect and insert into the wider intelligence jigsaw. Tour crews now were systematically mounting 'attacks' on airfields, targeting both aircraft, their weaponry and the missiles and radars protecting them. They were checking EDAs, into which Soviet forces would 'crash-out' defensively or aggressively as the immediate prelude to war, barracks installations, training areas, communications sites and bunkers, logging the movement of units by road and rail to and from exercise areas and collecting technical intelligence relating to Soviet armaments and equipment that revealed capability. By this stage of the Cold War, any pretence of amity or any vestige of allied solidarity was long over. BRIXMIS was engaged in its own war, a 'hot' intelligence war against a potentially deadly enemy. The perception was that any dropping of their guard might have resulted in the Western NATO alliance being caught lethally off guard, with catastrophic consequences.

To tour effectively as intelligence gatherers, crews were forced progressively to develop and evolve innovative tactics and techniques, sometimes displaying imagination and the application of lateral thought. A touring countermeasure adopted by the Soviets in the 1950s was the practice of surrounding installations with wooden stockades to frustrate both the casual observer and the deliberate 'stalker'. During the spring thaw after a harsh winter, the boards from which the palisades were constructed would shrink, creating regular and even gaps between them. Through trial and error it was discovered that by maintaining a constant speed of just below 50km/h, the equivalent of around 25mph, the spacing between the individual stakes merged into a unified 'window' revealing an unimpeded view inside. To tackle other high-value targets that were aggressively defended by intensive MfS surveillance, USMLM would deploy a number of tour teams simultaneously, ensuring that two or even three tour vehicles decoyed the surveillance teams away from the target whilst the 'real' team would mount the intelligence attack.

The riots centred in East Berlin in 1953 carried seeds of instability and miscalculation within them. How would the Russians respond? How would any response affect or impinge on the western alliance and specifically the status of West Berlin? It was one of those increases in tension and turns of the screw that ebbed and flowed throughout the duration of the Cold War. The Mission was not able to enter the east of the city to directly monitor events there, but it could watch the response as units of GSFG were mobilised from the surrounding area and moved in towards the capital. Tour reports supported a limited, controlled and restrained Soviet reaction. Elements of three divisions were observed moving into East Berlin, enough to project the impression of power but not disproportionately large enough to further escalate violence.

The year previously, air tours began to witness the arrival of the jet age in the DDR as old prop-driven aircraft were replaced by jet-engined MIGs. Perhaps more significantly, observation of Soviet exercise play suggested the Russians were not factoring nuclear exchanges into their exercise planning. This was a vitally important reflection of the Soviet perception of western intent and a reflection of their own.

The conduct of air tours mounted by the RAF contingent within the Mission began progressively to shape the conduct of all tours, assuming a more professional and systematic manner adapting flight planning methodology. The airmen led the way in 'professionalising' touring throughout the unit. In 1957, it was airmen who began photographing the targets they were observing, which up until then had been recorded in an ad hoc manner using sketches and other subjective methods of recording detail. This was the real game-changer in terms of touring tactics and maximising the quality and quantity of intelligence collected. Their approach of using observation posts from which to clandestinely observe Soviet activity became universally adopted. It was now SOP to ensure that *Stasi* surveillance had been evaded before teams continued on task.

The previous year saw the commencement of the Chipmunk OBERON flights, with an agreed frequency of three per week, flying at an altitude of 500ft. The Chipmunk crews now also began to photograph their targets, guaranteeing new profiles and further enhancing the quality of material gathered. This pattern and mode of touring, air and ground, would now stay largely unchanged until the very end, some thirty years later.

That same year, a tri-Mission approach to tour planning was adopted to ensure that a tour from one of the three Missions would categorically be in a position to detect any significant movement of troops that might herald the enduring bogeyman of a surprise attack by the Warsaw Pact. Additionally, a coordinated approach regarding the frequency of visits to targets was crafted to prevent tours from the other Missions becoming victims of enhanced security resulting from a previous tour that had perhaps visited a site only, conceivably, the day before. Intelligence reporting was also standardised on a tri-Mission basis, and within the British Mission an intelligence database was set up to record and ease retrieval of order of battle detail, the forerunner of ELECTRIC LIGHT that my Research office maintained.

Incidentally, paradoxically, this more joined-up approach had unfortunate, unforeseen consequences. It was highly likely that forecasts of upcoming tours were disseminated widely enough to be read by George Blake, a Secret Intelligence Service (SIS) officer sitting on the same corridor as BRIXMIS – the station later decanted to a suite of offices one floor down – as yet unidentified as a Soviet agent. Passing on the details to the MfS, via his KGB handlers, he would have enabled *Stasi* surveillance teams to impose effective surveillance more readily. Blake was only arrested in 1961.

As the 1950s drew to a close, BRIXMIS intelligence landed directly on the most important desk on the planet: the one dominating the Oval Office in Washington. The existence and nature of SA-2 radars in the DDR was marked up for the President's eyes. A congratulatory note made its way back to the Olympic Stadium in recognition. It was the SA-2 that shot down Gary Power's U-2, so it was deemed a weapons system of enormous strategic significance in these early days of intelligence-gathering overflights.

Arguably, one of the most significant events of the Cold War was the construction of the Wall in the summer of 1961. Politically, the immediate debate amongst Western leaders was how to respond to Ulbricht Doctrine, the SED leader's Soviet-endorsed initiative. The crux of the hotline conversations was, should NATO take any direct action in response, not least tasking a troop of local Berlin garrison tanks to bulldoze parts of the new construction? BRIXMIS had noted the build-up of Red Army forces to the west of the city in the local area around Potsdam, assessing them as a ready reserve to react to any NATO initiative. This reporting allowed cool heads to prevail, realising correctly the significance of this quick reaction force and the clear signal that the Soviets were almost certainly prepared to use it. This potential flashpoint was effectively de-escalated but the West's passivity ensured the Berlin Wall tacitly remained for its lifetime as an irrefutable symbol of east–west divide.

In 1962, the first T-62 main battle tanks were introduced into the DDR. The Mission confirmatory sightings signalled a critical enhancement in the Soviets' armoured warfare capability as the T-62 was armed with a new 115mm main armament, with its increased accuracy. The following year, T-55s were spotted river snorkelling. This coincided with initial sightings of the PMP pontoon bridge system in the country. Effective river crossing tactics and capability signified aggressive rather than defensive intent, greatly enhancing potential Soviet mobility.

On an equal footing with the construction of the Berlin Wall, the Cuban Missile Crisis, in terms of the scar it left on the collective memory, is recognised as the pre-eminent single event of the Cold War. Its consequence in eastern Germany, for all three Missions, was an internal realigning in operational priorities and output. Now, without any compromise, BRIXMIS devoted its efforts and resources to intelligence collection. Liaison was relegated even more emphatically to a distant secondary duty. The expansion of manpower that the mission now enjoyed concentrated even more diligently on observing and monitoring military activity, and disseminating the resultant intelligence – tension in one sphere of geographic interest naturally led to a ramping up of potential for conflict in all other realms of Cold War competition.

Having been on hand to monitor the transition of power from Stalin to Khrushchev, now with the latter also history, in response to an either thoughtless or intentionally provocative invitation for the Bonn parliament to sit in session in Berlin, BRIXMIS was now key in reporting on the Soviets' retaliatory second Berlin 'blockade'. A division was reported heading towards the IGB around Helmstedt. A whole airborne division para-dropped into one of the major training areas direct from its base inside the Soviet Union; all land access routes to the city were closed by the Russians, and aircraft, both civil and military, were buzzed as they transited the air corridors to Berlin. Yet another period of enhanced tension that required informed, cool political heads to manage it. That necessitated timely and accurate raw intelligence. The Missions were yet again ideally placed to contribute their share of the brushwork to the overall picture. When a BRIXMIS tour spotted the airborne division mobile and en route returning towards the eastern frontier with Poland, it was clear that, as in the Missile Crisis, peace had endured.

The consequence of the Mission reporting its first sighting of a Scud-B on DDR soil in 1965 was the whole balance between the forces in western Europe changing. To counterbalance the US positioning of Pershing nuclear-capable missiles there, the Soviets had now levelled the playing field. Like Pershing, Scud was a tactical battlefield nuclear-capable system that could also deliver a conventional explosive or chemical payload. Ultimately, this move probably contributed more to keeping and maintaining the peace, but knowledge of it was an important escalation with which governments were able to leaven their political and military calculations.

In 1966 a two-seater Yak-28 Brewer tactical bomber crashed into the Stössensee Lake in West Berlin. The aircraft had suffered catastrophic engine failure and the crew had valiantly attempted to divert to a Soviet airbase in the DDR before crashing. An intricate and well-coordinated recovery operation, undiscovered by the Russians at the time, plundered its jet engine technology and avionics. The operation proved to be a valuable barometric update, allowing Western scientists to gauge the real, not perceived or estimated, level of threat posed by Soviet aircraft. Only real knowledge can allow governments to accurately assess threats and respond accordingly, realistically and appropriately.

In the summer of 1968 the Missions' freedom of movement in the south-east of the DDR adjacent to the Czech border was impeded and restricted to motorway transit only with the imposition of three TRAs. Unable to see into the newly restricted areas, the Missions deployed to deduce what activity was being obscured by monitoring movements from barracks and garrisons in areas further north. BRIXMIS reported moves of units in the local area, both Soviet and NVA, heading south. The Mission's reports of well-disciplined and purposeful deployments were met with scepticism back in the zone but in the end, 400,000 Warsaw Pact troops were moved into Czechoslovakia, 75 per cent of them Soviet. Sightings of deploying equipment noted a new surveillance radar, Long Track, and a new height-finding radar, Thin Skin B. The immediate impact of the reaction to the Prague Spring was that the Missions, from July to November, found themselves under greatly enhanced *Stasi* surveillance as they continued touring, but continue they did with their vigour and tempo unbroken. Another crisis point had been successfully navigated and managed. Actions and responses had been measured and deemed proportional.*

An incredibly productive year followed: 1970. The SPG-9, a man-portable 73mm recoilless gun, was deployed to GSFG; SA-4 Ganef, a medium-to-high-altitude surface-to-air missile, with two warheads mounted on a tracked chassis and accompanying radar, was photographed; a new warhead was seen deployed on the missile of the short-range wheeled Frog-7 mobile single-rocket launcher; samples of new aircraft ammunition were collected from a live-fire aircraft range; the new air-to-ground missile fitted to the variable-sweep Sukhoi Su-17 Fitter fighter-bomber was photographed; and a new pod was noted on the MiG-21 Fishbed recce variant.† Soil samples were collected from a new runway, revealing important details of technical construction, and joint GSFG/NVA NBC warfare training was observed.

The following year, a Fitter was observed, by an air tour, troublingly engaged in a low-level nuclear air burst simulation. A Fishbed was seen

* This hardening of attitude on the part of the Soviets endured. Illustrating the tougher attitude to AMLM touring, between 1979 and 1983, BRIXMIS tours were rammed or detained a total of twenty-five times; there were four incidents where live fire was directed at the Mission's vehicles and another four occasions during which BRIXMIS tourers were beaten up. Over a slightly longer period, 1951 to 1985, USMLM tours were detained a total of fifty times.

† The MiG-21 is the most produced supersonic jet in aviation history and is still in service in the world sixty years after coming into service.

firing its 23mm canon and a J variant firing an air-to-ground missile for the first time.

In 1975 new touring equipment entered the BRIXMIS inventory. Tours were now routinely equipped with 1,000mm mirror lenses and external camera motor drives to enable bursts of continuous picture taking, and night-vision goggles (NVGs) facilitated covert night-time movement. Deployment of NVGs massively enhanced tours' capabilities. Not only could vehicles and individual tour members move around more safely and discreetly out of the vehicle during the hours of darkness, driving with them fitted allowed tours to travel at enhanced speed to shake off *Stasi* nocturnal surveillance.

By the end of the 1970s, Soviet forces, air and ground, had undergone comprehensive modernisation, exponentially increasing the potential threat posed to NATO, as usual under the watchful and systematic eyes of the Missions. This included the introduction of tracked artillery pieces, the 122mm 2S-1 and 152.4mm 2S-3; the sixteen-rocket 220mm BM-22 Multiple Rocket Launcher; more SA-6 Gainful (tracked with three missiles firing up to 50,000ft and out to 35 miles) and motor rifle and tank regiment organic SA-9 Gaskins (four short-range, low-altitude SAMs mounted on a wheeled BRDM-2 chassis). In the air an An-26 Curl aerial command post, DR-3 drones, mine laying Hip helicopters and Swatter anti-tank missile-fitted Hind-D helicopters all joined the arsenal. For the first time the three-man crewed 125mm T-72 tank was observed in East German hands.* A coordinated tri-Mission operation captured the first pictures of a Fitter G recce variant, and the arrival of the MiG-23 Flogger – the first Soviet fighter with a look-down/shoot-down radar, and armed with beyond-visual-range missiles – and the Hind-E in the DDR inventory. On the air side, close observation of the tactical use of *Autobahns* as auxiliary airfields was made. A number of new airfields were kept under periodic observation as building continued and key runway construction material was again forwarded to the specialists for analysis.

Geopolitically, the 1979 Soviet invasion of Afghanistan dominated contemporary Western political thinking. The Missions observed the deployment by air of troops from GSFG to bolster the invasion force. The war affected BRIXMIS in two direct ways. Firstly, in response to FCO

* The T-72 was never used by Soviet forces in the DDR.

directives, liaison with the Russians within the DDR was strictly curtailed and remained limited well into the 1980s. Social contact withered on the vine overnight. Secondly, the Missions received intelligence collection tasking related to the Afghan campaign, not least confirming the use by the Soviets of chemical weapons there.

In tandem, the political situation in Poland once again required BRIXMIS to monitor any nascent Moscow-led invasion plan on this, their 'western' front. With TRAs obscuring the view of the Polish frontier and very limited opportunities to quiz SERB and other Russian contacts, post-Afghan adventure, it was difficult to assess if the situation within Poland would deteriorate to any dangerous new degree. Eventually the imposition of martial law by the Polish military government stabilised the situation.

The new decade began with two scoops, which helped redress the balance, when BRIXMIS recovered examples of the new 5.45mm round for the new AK-74 assault rifle, and the low-tech application of a tungsten-tipped scraper revealed the composition of the new T-64 MBT armour. The new BTR-70 wheeled APC was recorded for the first time sporting GSFG VRNs.

Always of especial importance was the identification of new communications-related equipment. The Twin Ear radio relay and tropo-scatter system was spotted, as was the electronic warfare (EW) Turn series and a new jammer, plus a new satellite communications system with the NATO nomenclature of Wood Bine.

The first year of the 1980s continued with routine rubbishing discovering a discarded notebook in which technical details of new Soviet tank armour was recorded. A new nuclear-capable 152mm turret-less SP gun, the 2S-5 first used in combat in Afghanistan; the RKhM tracked NBC contamination monitoring vehicle and the tracked SA-13 Gopher, a short-range, predominantly anti-helicopter, air defence vehicle, were all added to the scorecard. Of great significance, reinforcing the perception of the Mission's 'reach' as well as the quality of its intelligence product, a BRIXMIS officer pulled off a stunning coup, gaining entry to the turret of a T-64 tank parked up in a shed and photographing all its systems.[†] Additionally, a new variant fitted with a laser range finder was photographed.

---

† Soon after, he left the Army and moved on to government pastures greener. We worked together on some joint projects in Cyprus.

The following year, 1981, clandestine entry into a Soviet nuclear bunker and the removal of a filtration unit provided an insight into Soviet NBC research. A nuclear warfare command train was observed, and a technical report of an 'attack' against a tour using a laser was disseminated, which enabled subsequent analysis to provide valuable detail. A complete AT-7 94mm wire-guided portable anti-tank system, comparable to the UK's Milan, was incredibly recovered to Berlin without compromise. But the most important reporting revealed the Soviets had put into the field the short-range amphibious SS-21 nuclear-capable battlefield missile system. New aircraft variants to bolster GSFG's air capability included the supersonic MiG-25 Foxbat, capable of flying operationally at Mach 2.83, Fencer Cs and Fitter Ks.

With the arrival of a new commander-in-chief of GSFG in 1983, the tough veteran General Zaitsev, the Cold War, as it affected the European mainland, cooled yet further, increasing the pressure on the Missions to provide vital insights into the level of preparedness of Soviet forces inside East Germany. This was made more difficult by the increased imposition of PRAs. Zaitsev was old school and dangerously shared the belief of the current Moscow leadership that the West was capable physically and mentally of launching a first strike against the Communist Bloc. He worked ruthlessly to further harden his command as an efficient and effective fighting force driven onwards by this misapprehension.

Despite this, air tours were able to evaluate Soviet flying performance as generally lacking flare and dismissed their low-flying skills as poor – vital intelligence and further informing the appreciation that the Reds were not the tabloid multi-headed hydra. However, concurrent direct observation of helicopter training debunked the conventional wisdom that Soviet front-line helicopters could not operate in severe winter weather. This had an obvious implication for assessing Soviet tactics in the mountains of Afghanistan and precipitated a reassessment of the West's counter-strategy closer to home.

The new M1976 152mm towed howitzer was spotted in a rail siding and duly comprehensively photographed, and the new BMP-2 main gun calibrated. The new gas-turbined 125mm main gun T-80 main battle tank was identified in-country, initially under tarpaulin, and a portable pontoon bridge – critical equipment if Warsaw Pact forces were to invade the West – the PMM-2 ferry bridge based on the T-64 tank chassis, was spotted.

Enhancing the importance of the Hind, an F variant was identified with a direct depth anti-armour role, rather than the established mission of providing close air support to infantry forces.

The year 1984 brought in none of the dystopian drama predicted by George Orwell but the confirmation of the widespread equipping of GSFG's MBTs with ERA reinforced the Western perception of aggressive intent. ERA dramatically improved the tank's protection from anti-tank missiles and opposing MBT main armament rounds, as did the appliqué armour retro-fitted to BMP-2 APCs, equipment critical to the Soviet doctrine of fast-moving *blitzkrieg*-style assault. First sightings were made of 'naked' T-80s at night moving by train – these were the first full-frame, comprehensive, all-body shots. The K-611 NBC detection vehicle was photographed for the first time in eastern Germany, and the Sukhoi Su-25 Frogfoot ground-attack aircraft's arrival in the DDR was confirmed. An airborne jammer version of the Hip helicopter was spotted in tandem with the unexpected, and unexplained, sighting of a Hind-G1 fitted with a nuclear testing scoop.*

Gorbachev's 'coronation', as we would soon see, was to shape the Missions' destiny more radically than any other event in recent Soviet political history. Very quickly a sense emerged that things would never be the same again, but just how revolutionary his approach to domestic and foreign policy would be eluded all the West's Kremlinologists. The Zaitsev restrictions on Mission movement lapsed, and contact with the Soviets increased, resuming their pre-Afghan invasion levels. In response to Gorbachev's political overtures to the West, BRIXMIS's touring profile was adjusted and became less aggressive, less potentially provocative and more 'stand-offish'. Targets were tackled from a distance with a large lens wherever possible. This is not to say, however, that tours did not operate with the aim of gathering as much intelligence as they could, nor with 'vigour' when the tactical situation demanded it. But, looking slightly ahead, with the withdrawal of 80,000 GSFG troops in 1988 and a clear statement that there would be no prospect of a Soviet first strike or any offensive ground force action in western Europe, at least with a degree of hindsight, it was already beginning to be clear that the Mission's twilight years were advancing rapidly and might themselves not be long lasting.

* It was seen in action in the immediate aftermath of the Chernobyl disaster.

A number of scoops marked 1985's touring. First, photographs of the Antonov An-26 transport aircraft fitted with external bombs, a modification to the SA-2 Fan Song air defence radar, enabling the system to record images of inbound enemy aircraft, and highly detailed technical photographs of the Tin Shield radar, part of the SA-10 system, were processed and disseminated. Mission tours continued to reveal more detail of the deployment of T-80s in GSFG. Complementing the intelligence collected detailing ERA fitted to a T-64, it was confirmed, unsurprisingly, that the new T-80s were also 'hardened' with explosive armour blocks. Right at the top of collection priorities was any intelligence related to NBC equipment, aggressive or defensive in role. A new mark of respirator and a chemical protection canister issued to foot soldiers were picked up and brought back to West Berlin. Two new SA-5 Gammon sites were also recorded and Fitter G's deployment at the new airfield at Laage, on the north coast, confirmed.*

The highlight of Mission collection for 1986 was the find of a document providing a comprehensive summary of the T-80's technical characteristics, and the first close-up photography of the tank itself. The recovered document listed everything from deployment range to main armament rounds and anti-tank missiles carried to fuel consumption. Other scoops included sightings of the MiG-25 Foxbat F; the MiG-29 Fulcrum air superiority fighter designed to counter US F-15 and F-16s; the SA-11, a very versatile medium-range system; the 2S-9, an air-droppable 120mm gun-mortar designed to support airborne forces; a radar variant, plus command and control versions of the multi-variant MT-LBu; variants of the PRP-3 recce vehicle and confirmation of the coordinated way in which they were integrated; and definitive confirmation of the armaments deployed by the tracked 2S-6 with its unconventional mix of anti-aircraft guns and missiles.

The scoop of 1986 was the recovery of an intact block of T-80 ERA. The following year then continued with more strategic-level intelligence with the first photographs of the mobile nuclear-capable battlefield-deployed SS-23 missile in the DDR, designed to replace the Scud-B, and reporting of short-range SS-21 deployment down to brigade level. Initial

---

* The SA-5 was a long-range, medium-to-high-altitude surface-to-air-missile designed to target incoming enemy bombers.

photographic evidence of the deployment of over twenty variants of EW vehicles was recorded including both collection and jamming capabilities. A number of TM-62 anti-tank mines were transported to the Stadium in the back of a tour G-Wagen for onward despatch to the government technical research facilities.†

The hand of Gorbachev was clearly in evidence when, in 1988, BRIXMIS recorded SS-23s based in the south of the country withdrawing back to Russia without any prior public statement. Despite Gorbachev's downsizing of GSFG and the re-balanced relationship with the west, new equipment did, however, continue to enter service in-country. Scoops for BRIXMIS included thermal imaging signatures of main equipment, a task undertaken by BRIXMIS SF tourers; photos of the new Beriev A-50 Mainstay AWACS; T-80 snorkelling; a new recce pod fitted to the Flogger J; enhancing GSFG's command and control capability, the identification of twenty-plus variants of Silver Box mobile HQ vehicles and the first photos of the Copper Log computer vehicle.

With absolutely no sense of its own impending demise, nor the upending of the post-war status quo that was but months away at the time, BRIXMIS's final significant scoop came in September 1989 at Storkow. It was that recording of the first sighting of an SA-10 Gecko in East Germany – where GSFG had been re-designated as Western Group of Forces – and caused the last great stir in the Germany-focussed tech int community. Another great job well done.

† When a US crew returned home with live munitions on board, they were roasted by their Chief and threatened with the sack. Personally, I think this approach was probably much more sensible than a tacit acceptance of a potentially very hazardous exercise. USMLM tourers were thenceforth ordered to photograph and measure unexploded ordinance but to leave it categorically in situ.

13

The New Normal

As things rapidly opened up and the new normal arrived and seamlessly blistered on to, rapidly beginning to reshape, eastern life and society, I had to conclude what a monstrous waste the whole DDR experiment had been: what a monstrous waste of opportunity, wealth and every other metric and measure that could be used to assess and value the previous forty odd years of the state's existence, not to disregard the twenty-seven years that the Wall had segregated the eastern *Länder,* whilst incarcerating its people.

But it was also sad to see the initial tsunami of euphoria on both sides of the Wall dash destructively on the rocks of cynicism and greed. Unscrupulous sharks from the west rushed headlong into the new states of the east, dumping forecourt pups that would not sell to more discerning western car buyers, and villainous carpet-baggers who moved, locust-like, stripping art, antiques and anything else of value that they could remove for a few fistfuls of hard west marks, carrying them back lustfully to the west. In reaction to the collapse of eastern economic enterprise, the resultant soaring unemployment and sub-West German salaries, an increasing section of former DDR citizens began steadily to bemoan their new-found freedom and expressed remorse that *their* Wall was gone – *Die Wende*, the change, was one for the worse not one for the better in their smarting eyes. This was compounded by negative sentiment in the west as taxes increased, despite Helmut Kohl's emphatic promise to the contrary, to pay for dragging the east forty years forwards, into the 1990s. Sad but probably inevitable. The rest of Europe now looks on as a unified Germany dominates the continent politically and economically, despite its current issues.

In my opinion, reunification did not have to have been the unchallengeable certainty that it quickly appeared to be as 1989 moved into 1990. Eastern Germany could have remained a sovereign nation under a democratically elected government. With adequate internal economic reform and enough external support from the EU; the US, with an updated Marshall-type plan; and other agencies, it could have evolved, slowly and painfully, but positively, into something new, bright and enduring instead of succumbing to the new *Anschluss* that engulfed it and swept aside anything and everything that had 'East' or 'Socialist' stamped on it. It would have been a troubled road and certainly not easily managed, but it was, for me, worth at least the effort. Kohl was motivated by one thing, and one thing only, to the exclusion of all else. He wanted to be the new Bismarck: the chancellor who re-made Germany – all-devouring hubris. I thought at the time, and still do, that for him no price was too high to pay for this laurel wreath.

At the micro level, life for the Missions initially carried on superficially unchanged, but to me, and I'm sure for others, it did so against what felt, very rapidly, like an increasingly artificial backdrop and an increasingly incongruous coexistence with the changing face of the old DDR and, indeed, without overstating the issue, the world as a whole.

Glienicke Bridge exemplified and epitomised this perfectly and succinctly. Of course we still crossed into Potsdam via this time-honoured route, but we now shared the privilege with the steady stream of civilian traffic that flowed in both directions. It was a bizarre precedent. We stopped to hand over our passes whilst they swept past us as the pace of western life now washed backwards and forwards into Potsdam and the rest of the east.

Christmas came and went with the hardening perspective that the eventual outcome of *Die Wende* would be the reunification of Germany and the demise and complete disappearance of the DDR – all its institutions, organisations, responsibilities, treaties and icons, good and bad. Surely, clearly, this would see both WGF and the Allied presence vacating the scene, and with them, naturally, the Missions too – choppy waters ahead. This lack of coherent clarity concerning the future very quickly began to affect morale, certainly within the British Mission. The one thing that soldiers really hate, more than any prospect of physical danger, is uncertainty. Soldiers want to know what tomorrow will bring even if the day will be dark, cold and hazardous. Everyone in the Mission was bright, recognised the forces at play and understood the implications of the political factors unfolding around them. The only real question that had to be answered was how long the uncertainty would drag on, and for how long the life of the Missions could be extended, a bit like anticipating the final squeeze of an emptying tube of toothpaste. For those with families, the sense of a rudderless ship beginning to flounder was even more acute. The tourers who arrived from the next, and final, course found themselves in an even more invidious position. The Chief had a job on his hands managing not just expectations but maintaining the operational focus of the unit. I have to say that he rose to the challenge and I doubt many other senior officers could have led the Mission as he did. He was trusted by his men and women.

New Year's Eve, however, successfully dulled any nagging concerns and worries about the future, at least temporarily. A huge street party again pulsed and throbbed in the area around the Brandenburg Gate and the Reichstag. It felt as though not just the whole of the city, east and west, but half of Europe had gatecrashed the bash. Fireworks rocketed, spun and exploded everywhere, way too many shooting dangerously horizontally at head height, like mini RGP-7 rockets. There was none of the control and health and safety concerns of a British Guy Fawkes Night.

A group of us had started the evening at the Chief's residence for drinks, then moved downtown to join the party. East mixed with west in what was now a completely unremarkable common reel as we mingled and pushed through the crowds. At some point I became separated from the BRIXMIS band and fell in with a group of revellers from over the Wall. As midnight tolled and 1990 dawned, they tried to persuade me to return the other way with them to carry on partying. I considered it for a second and kindly declined. Paperwork was still being checked at at least some crossing points against an ad hoc and unsystematic new 'border' regime – to be stopped and questioned could have led to a difficult and embarrassing situation, and ultimately to that early flight home if things accelerated out of control and went really wrong. It was, I think on reflection, a sound decision despite the fact it would have been great fun and a unique experience.

The following afternoon a couple of friends and colleagues from the Mission and I had decided to take part in the 6km run that would historically and symbolically breach the Brandenburg Gate, *the other way*, head east and circle the historic centre of East Berlin. To shrug off the weight of the still erect, but by no means intact, Wall and tread on these iconic pavements that I had only been able to observe from the air and from the tourist platforms, was an exciting piece of history for me. These really were the streets of le Carré and Deighton. Ed, from my 'near-death' electric tour, Gary and Mal – a Tornado pilot and full-time tourer respectively – and I joined the 25,000 other runners who had crowded together in good humoured, expectant anarchy at the start point in the Tiergarten park. Maybe there was a gun, I don't remember, but eventually we were mobile, winding our way around the course.

As we passed through the gap engineered in the Wall's slab face, it was a strange yet exhilarating feeling. We were going against the tide even then, but to be passing right under and through the Gate where, just two months ago, history had been frozen with absolutely no prospect of thaw was just not computable. We were treading where no one but the border guards of the now-dead regime had trodden since 1961. It was my first, but certainly not my last, visit to the eastern side of the city. It could have been a marathon and I would not have struggled as I rubbernecked the whole way taking in the sights. If the run was construed as a race, we were not participants.

We passed the Soviet embassy on Unter den Linden, under the lime trees, whose inhabitants must have been sharing the same sense of uncertainty and doubt, not only about their personal fates, as we were experiencing, but the wider future of their state. I picked out *Neue Wache* with its distinctive half a dozen Doric columns on the left, with its pair of jack-booted NVA sentries. The regime had re-dedicated the building with its eternal flame to the victims of fascism and militarism. The guards were still on duty. I have always puzzled why the dress uniforms of the *Nationale Volksarmee* so closely resembled those of the *Wehrmacht*. Surely the party must have appreciated the similarity, yet why not demonstrably create clear blue water between their army and the forces of Nazism? It could not have been an easier 'win'. In my mind the resemblance reinforced the totalitarian similarities of the older and now old state and underscored the inherent shared principle of German militarism.

There was the impressive home of the Humboldt University, the eastern half of the University of Berlin once it split with the division of the city. Past students included Bismarck and both Marx and Engels, and Einstein had been a professor there. As we drew level with the university, we could clearly see the base of the distinctive *Fernsehturm* TV tower lofting 360m above Alexanderplatz, and the baroque dome of the protestant *Dom*, the largest cathedral in Germany. We passed, on our right, the East German parliament building, the *Palast der Republik*, a modernist structure completed in 1976 with less than a year of life left, before we left Museum Island. We crossed over the River Spree, a silent witness to a number of intrepid escaping swimmers traversing its freezing course over the years. We went as far as the High Renaissance *Rotes Rathaus*, rebuilt in the late 1950s after being virtually destroyed during the Second World War, circled the TV tower and retraced our steps, with no loss of bounce, I have to say. It had been a totally exhilarating experience. It was a brilliant city tour, to us, feeling very much as though we were taking the first cautious steps of intrepid explorers in a fabled new land, and what a wonderful start to the new year, the Missions' last.

I think it was Gary's idea: he provided the photos and I wrote an article for one of the UK's major running magazines and we sent the article in for publication – to my surprise, it hit the news stands. It was certainly a historic run, but wider than this, it was an event of some historical significance in its own right.

14

Russian Friends

The interface between the Missions and HQ GSFG/WGF in Zossen Wünsdorf had always been, right from those initial years, and continued to be right up to the end, an interesting organisation that went by the name of SERB. At that stage, in modern British military history, the name had none of the lasting wider associations that would very much follow half a decade later, in the Balkans. Headquartered, logically, in Potsdam, SERB's role was two-headed. Overtly, it represented the liaison wheels and cogs that ensured all the logistical and administrative elements underpinning our existence in the east functioned smoothly, and that when we needed help in contact with any Soviet 'agency', we got it. They were the channel through which protests were passed detailing our bad behaviour and through which we protested formally when unjust deeds were done to us. Where conflict arose over a contested incident, it was SERB's role to investigate and inform the command. Their more important role, however, to their political and military masters, was to provide another intelligence collection channel reporting on Mission personnel and activities.

To this end, a substantial proportion of the dozen officers – in my time – whose existence we were aware of, were assessed as being military intelligence officers from the GRU.*

* As detailed in Annex A, the *Stasi* officer I 'ran' commented that there was always a lone KGB officer serving within the SERB team, no doubt to ensure political solidarity and discipline was maintained at all times. His identity was not readily apparent to MfS officers of 'Frank's' seniority, and I remain unsure who worked for the Lubyanka in my time.

Of these, the characters I remember most vividly were the deputy head, Colonel Robert Polozov; Lieutenant Colonel Yuri Pliev, a larger-than-life South Ossetian and self-proclaimed fan of Shakespeare's works who, so the story goes, worked on North Sea 'trawlers' in his early career (I doubt as a fisherman, rather, I suspect, in a GRU signals intelligence collection role); Major Sergei Savchenko, a Ukrainian who spoke near faultless English with a pronounced American accent who had conducted the SERB investigation into the fatal Nicholson shooting; and Major Sergei Ostroumov, with whom I established the strongest personal link. They possessed the self-confident openness of the HUMINT officer in contrast to the introversion and caution of the 'regular' Soviet we encountered. I remained unsure about the very pale Lieutenant Colonel Demchenko. I did not assess him as being an int man, and knew he had attended the Chernobyl disaster in the first five days in his role as a military engineer. We also regularly rubbed shoulders with Captain Dimitri Trenin, who was, I think, a full-time interpreter – I have seen him a number of times since, commenting on Russian affairs for the BBC. The chief, Colonel Pereverez, was a tall, urbane, poised, grey-haired officer – I do remember him talking about his childhood in the Great Patriotic War and his morbid confidence that both his and his wife's lives would doubtless be shorter than the average and probably plagued by ill health as they aged due to the extreme privations of the war. These stoically made comments have always stayed with me. I have pondered too if Demchenko's heroism cost him a premature death.

I think one of the key lessons from the Cold War was the Western failure, even though we claim victory and award defeat to the Soviets, particularly on the American political side, to really understand and get inside the Russian psyche. I still feel this is certainly a prevalent failure in the way we collectively deal with Putin's contemporary Russia. Certainly, similar intelligence failures in reacting to other situations and informing initiative to guide developments have been the consequence of flawed or inadequate appreciation of fundamental 'ground truths'. These failures have been wider and are by no means limited to immediate post-Cold War Russian relations. Without question, they have led to our lack of success in both Iraq and Afghanistan. In both theatres we just have not understood – and do not understand – the key drivers, societally, historically and psychologically. These failures are certainly not new but have led to painful and expensive compromises in terms of output and outcomes, and had already characterised the reality of our handling of Cold War issues. In the context of this struggle the objective reality attests to a more accurate conclusion, I believe: the West did not win the Cold War; we just did not lose it. Ultimately, the Soviet Bloc collapsed due to exhaustion, a lack of confidence at the top and unrealistic overspending on perpetuating its military prowess. The apocryphal story of a dead Brezhnev being physically propped up to continue running the regime is apt. The Soviet Union died and was being artificially maintained on life support long after its economic failures had bled the corpse to death. To me, looking back over the span of decades, it is a miracle that the lack of mutual understanding and the degree of underlying distrust did not degenerate into direct armed conflict. I am positive the Missions played more than a bit part in preventing this catastrophe.

We would regularly socialise together with SERB officers in the Mission House at official high-profile and more informal functions, and I remember on one occasion we were invited to their 'club' located within their office building in Potsdam. It was a very special evening. The Soviet wives would sometimes accompany their husbands – this was one of those occasions. Each couple brought in dishes from home that they had sourced and prepared to create a really substantial buffet of Russian food that could not have been more authentic – there was no party service cheat in those days in Potsdam. There were *blinis*, rich *borshch*, sour cucumbers, salads, carp, caviar, plate after plate. It was a deeply personal gesture on the part of our hosts and certainly touched an emotional chord in me.* As the evening wore on, inevitably a guitar was produced and those who fancied themselves as minstrels, with or without the vodka that flowed freely, chimed in. Those who were really short of a grasp on reality even attempted to keep pace drinking with our hosts – the straight, but short, road to destruction. Of course a key component to any evening in Russian company was the endless toasting as the reason and justification for downing each shot. I remember few evenings of such warmth, humour and just good honest camaraderie.

* Up until then, my experience of Soviet food had been the rations that we regularly took home. As we were administered by GSFG/WGF, this responsibility and duty included feeding us, notionally, in reciprocity for the rations we dispensed to SOXMIS in the zone. I will never forget the joints of pork from pigs farmed on the edge of every military garrison. Butchered differently from our approach in the West, the meat was red, and so strong was the smell that I could not get it past my nose, and I am certainly not a fussy carnivore.

Of course behind the bonhomie no doubt the intelligence machine continued to function and reports were compiled the next morning to update the files that must have been kept on all Mission personnel – I know that our post-evening actions were no different.*

1988 had ended with a New Year's Eve party, also hosted by SERB. Very cheekily, but marking their humour, they had covered the walls of their anteroom with a skilfully rendered series of cartoons, drawn by whoever was the in-house Kandinsky, on butcher's paper. The cartoons depicted scenes from BRIXMIS tours during the year, some of them very near the knuckle and clearly based on very specific intelligence collection targeted at the Mission, by either the *Stasi* or the Soviets themselves. Sadly, I have not seen any photographs, if any were taken, but they would have reinforced the fact that the Russians knew a lot more about detailed Mission operations than we would have liked, and underscored their cheeky humour. As the evening continued, there was, of course, more and more vodka consumed leading to some shirtless table dancing by the two Chiefs, but in this sort of large group setting, innocent and harmless.

I am certain that SERB would have been assessing us for our potential as targets for recruitment. It would have been a long game and a patient process of building a profile for the future. I am doubtful that any approach to any target would have been made directly by SERB; this would have had the potential to provoke a significant diplomatic complaint if the approach failed – more likely, I am confident the venue selected would be at another time in another place. However, this did not prevent a direct recruitment approach being made to a USMLM warrant officer some years previously.†

---

* Years later, on my final posting in the Army, I asked a friend and colleague working for the Security Service-administered BSSO(G), the organisation in Germany that liaised with the German federal and state security services, to track down my *Stasi* file from their former Normannenstrasse HQ in East Berlin. It was brief, recording that I was traced as a military intelligence officer but nothing else – a testament to either my professionalism or my complete lack of importance. I have not been able to see any notes or comments that our Russian friends recorded. I would expect, however, that these were somewhat more comprehensive regardless of any conclusion drawn.

† As detailed in Annex A, the MfS had identified that the warrant officer managing the USMLM House on Lake Lehnitz in Neu Fahrland, a slightly more distant suburb of Potsdam, had developed an extensive business network, presumably in part, of dubious legality. The information was passed to the Soviets, who believed they could exploit the situation. A recruitment operation was planned but little did the Russians know their intent to turn the target was fully known to the Americans. Two Soviet intelligence officers were 'apprehended' in due course as they approached their quarry.

Equally, we worked to build rapport with our SERB counterparts. BRIXMIS would certainly not have taken any routine active role in any exploitative operation but we did certainly serve as 'eyes and ears' and provided an overt and highly visible channel through which a wavering Russian officer could begin taking soundings on changing sides.

My particular personal 'target' was Sergei Ostroumov, but I also made it my business to pass anything of interest concerning any of our contacts. Ostroumov was badged as an artillery officer, spoke good English and came from Moscow. He struck me as the contemporary 'oligarch', probably from family with party connections, slightly aloof, slightly arrogant, but quietly calm and personable. He did not strike me as handler material, so I remained unsure as to his role or responsibility. His wife, Nina, was a very pretty woman who had trained and worked as an underground railway engineer – so very Soviet. Sergei and I met a number of times informally for social evenings out and, after the Wall came down, usually around Potsdam, but once in East Berlin.

On this occasion, the venue was suggested by Ostroumov: East Berlin for a change. Breaking the routine of meeting on 'home turf' immediately made this get-together different. I could only speculate if this signified something significant. Or did it fit some more mundane parameter? We had agreed to meet under the elaborate giant twenty-four-sided Urania World Clock on Alexanderplatz. Why the communist authorities required a 16-ton timepiece is a moot point, but nevertheless it was an RV point laden with Cold War symbolism and drama since its erection in 1969. As we were meeting in civilian clothes, I chose to travel to Alex by *U-Bahn* and took the loop that I had mistakenly jumped onto in error nearly a decade before on my post-university attachment. The Berlin of my *now* was a totally changed city from those days *then*. Then, I had sat in my seat in a state of extreme trepidation waiting for the *VoPos* to enter the carriage and conduct a check of ID papers before the underground train looped back into the west. Now, I stood, just another passenger, and jostled my way out onto the platform and up into the fading light of the very epicentre of the former capital of the DDR. Not a *VoPo* in sight. Berlin was truly a newly reborn city in a newly reborn country, but there was the clock, reassuringly unchanged, a constant in time.

I was deliberately early and took up position about 50 or 60m away with a clear uninterrupted, but discreet, view, resting nicely in the thickening shadows. This allowed me to survey the ebb and flow of the strollers and loafers, idling away the evening, and the scurrying determination of workers making their way home more purposefully and speedily. As I people-watched, I was again met by the unassailable impression that the population in the east carried with them, on them, some inherent *differentness*, almost indefinable but no less real. They were indistinguishable in terms of features and build from their neighbours across the Wall – they were, after all, of the same genetic stock. But there was a pallor, a paleness, not exactly the look of a sickening people, more a lack of vitality, of spark, of spontaneity. As I looked on, I always felt that underpinning all the human emotions they shared and displayed in common with their western cousins, the people of the DDR were essentially inveterately cautious and inward looking. They were somehow more rooted to the here and now. As citizens of their state, this came as no surprise. Everyone had to find a way to survive, more difficult if you were ambivalent to your state, as I never doubted so many comrades were – even harder if you were opposed to it. This sense of reticence and unease cloaked their every movement as they went about their business. The clothes they wore marked them out collectively just as graphically. They were clean, they were tidy but they were different. It took a moment's consideration to articulate this. There was essentially a lack of style, certainly a lack of exuberance and the desire to make any form of statement – unless paradoxically and perversely you were an officially tolerated punk. They imparted the absence of impact. Colours were muted, or oddly tinged; cuts were utilitarian and practical rather than seeking to create impression or flatter. Fabrics looked unnatural and were largely the creation of man rather than nature. And there was that unique patterned, coloured blue denim again. I have seen it nowhere else and at no time since. The overall effect was a population *clothed* but not *dressed* – tawdry, mean and dispiriting.

Alexanderplatz is certainly not pretty but it is impressive and could only be the product of socialist central planning. I looked up at the TV tower with its revolving restaurant, then decidedly stationary. It was an ironic symbol on two counts. Despite never winning the propaganda battle of the two Berlins, it was the east's tower that dominated the city skyline, whichever direction one looked in or from. But its mastery over the western horizon was compromised by the massive cross dominating the glass sphere atop the tower, created by reflected sunlight on both a winter's and a summer's day. It was no longer visible now as the sun had dipped down behind me but one had to smile. That simple manifestation of natural science must have served as a continual irksome source of irritation to the SED leadership – ultimately, the creator and mother nature were against them, as history had so precipitously, uncompromisingly and unstoppably proved.

I checked my watch. I was looking for any signs that Sergei was not alone and had brought, either willingly or unwillingly, professional company along for the ride. All appeared normal as I scanned the square. I could not spot anyone amidst the thinning throng who looked out of place or conspicuous. No one appeared to be reversing my scan, looking intently at it or the area of artificial light now encompassing it. Just a few minutes late, I picked up Sergei making for our RV point below the clock in his brown checked tweed jacket, significantly with none of the stealth or taut balance of the hunter. It looked like he was relaxed and alone, not that the presence of any minder or minders would have deterred me from meeting him on that particular evening. But this absence of company reinforced in my mind his status as a trustee of his regime. I moved to join him, after another brisk half minute of informal surveillance, leaving the cover of my position and heading to where he was standing directly under the massive time piece in literal observance of the plan we had agreed upon. His greeting was effusive and positive and we moved off to a restaurant, which he chose, a brief walk away.

This meet, like all our others, was certainly water-testing contact. The water remained tepid, but at least relatively clear and pure. He gave me no signals, overt or more restrained, that he wanted to create any form of channel to Western intelligence – nor I the reverse. I believe these meetings, from the Soviet perspective, were speculatively to identify any weakness in my character or lifestyle that they could exploit at a later date. But more than this, realistically the meetings were a subject from which to create a report to demonstrate to the 'system' that SERB, and more particularly Major Sergei Ostroumov, was actively doing its job of monitoring not just 'ordinary' Mission personnel but in this instance a member of British military intelligence.

Nothing came directly from these liaisons with Ostroumov, or any others, except three items of strategic-level tittle-tattle that I reported: the military officer class loathed Gorbachev and held him responsible for unforgivably frittering away the Soviet empire and squandering its prestige and stature as a superpower. Secondly, there was real concern and anxiety about the future in terms of job longevity, and about living conditions back in Russia as WGF returned home – perhaps useful leverage in any future nascent HUMINT relationships with Red Army officers. Thirdly, the collapse of the Wall had been just as huge a seismic event and just as shocking and unanticipated to the Soviets as it had been to us. There had been absolutely no discernible preamble or indicators that its end was nigh. These points suggested that things did not bode well for a smooth future, politically, back in Moscow. I put pen to paper and in turn fed our 'system' – the pursuit of intelligence differs little, I suspect, the world over.

Beyond this, these meetings were certainly interesting as much as they were enjoyable and professionally noteworthy. My predominant memory is of the decor of the eastern bars and restaurants we met in. I am left with an overriding sense of 'brownness' set against the broader background 'greyness' – brown lighting, brown furniture, brown tablecloths and, of course, Sergei's brown tweed. Living with lignite. From the standpoint of a professional intelligence officer, the opportunity to mix openly and freely with members of the sinister state intelligence services of our pre-eminent enemy, arguably the most aggressive intelligence agencies in the world, and such powerful players in the Soviet system, was something very special indeed. These were opportunities that very few members of any of the Allied agencies could boast to have had the privilege to sample, and very specifically, so few members of my own Corps. These meetings were professional, safe 'meets', in sharp and graphic contrast to those I had engaged in in Northern Ireland, where the prospect of physical danger loomed large, but no less a tremendous opportunity to use all the professional lessons we learned as HUMINTers – just fascinating and such rare personal experiences on top of all the other unique opportunities life in the Missions generated.

My path crossed that of another leopard with some subtly different spots. He was another major by the name of 'Steve' Maltsev – at least that was how he called and introduced himself. I quickly ascertained that he had worked for SERB in the early to mid-1980s but seemed to have moved on, although was still based in the Potsdam area. He was clearly an intelligence officer and a HUMINTer whom I assessed was engaged as an agent handler in targeting operations against Allied military personnel, if not a wider span of targets. He attended a liaison function at the House; I don't remember the occasion or when, although it was certainly post-Wall, and that was where I first met him. I have to say I do not remember if he was introduced to me by someone else, SERB or other, or if he introduced himself. We got talking. He was easy to talk to; his English was accented but fluent. He was certainly confident, and we got on well, establishing an easy but, by its very nature, superficial and disingenuous rapport. Without effort or thought, we naturally slipped into an immediate agreement to meet again, by doing so of course confirming both our true professions – everything seamlessly scripted by central casting.

We met several times separately after this initial 'bump', including meeting up in the centre of East Berlin in the main bar of the thirty-seven-floor Hotel Stadt Berlin, now the Radisson Park Inn, off Alexanderplatz. The Russians did seem to favour very accessible and easily identifiable meeting points. This was probably deliberate policy, I suspect at least partly founded on the desire to enjoy the protection afforded by a very public place that would deter any nefarious acts by Western intelligence. Considering the other side of the coin, it was as good for me as it was for him. On this occasion I decided to take Martin with me.

We were, both parties, undoubtedly feeling out the other and looking for ways and means to advance, if this was possible. It is quite feasible that he was conforming to the Soviet modus operandi and arranging routine meetings with potential, in reality highly speculative, targets to meet a quota and generate reporting, even if it lacked any real punchline. Maybe his accounts of our meetings included little snippets he manufactured in order to spice things up. I have no way of knowing – obviously my reporting was scrupulously accurate with its institutionalised clear division between fact, opinion and speculation.* He certainly was not 'aggressive' in his approach. Again, nothing substantive, from my perspective, came from our contact before we mutually drifted apart, but there are two interesting postscripts.

Once BRIXMIS had been disestablished on 31 December 1990, and JIS (Berlin) had assumed responsibility for monitoring the now rapidly withdrawing Soviet forces from the now former DDR, we were far more

* In writing reports reflecting human source meets, we scrupulously observe the distinction between the information that the source gives us and any interpretation or context that we append to it. This is clearly marked as 'Comment'. Thoughtful Comment adds value to the report because it reflects the experience and knowledge of the handler without misrepresenting the source's information.

mobile and aggressive in our touring routines. Befitting this, I went to attend a Soviet airbase open day, joining the throngs of local people who had taken the opportunity for a day out on formerly forbidden territory – incidentally, I also gatecrashed a similar event in East Berlin at the Karlshorst HQ, where the KGB had a large unit stationed to which Putin regularly reported from his Dresden office. As I wandered amongst the crowd, moving from exhibit to exhibit, including a really very good rock band comprising Red Army soldiers complete with blonde female lead singer, and a functioning field kitchen feeding visitors typical Soviet Army fare, who should I spot amongst the mass of thousands and thousands of day-trippers, also in plain clothes, but 'Steve' Maltsev. His bright-blond hair, ruddy complexion and slight paunch were not hard to recognise amongst the colour and commotion of the weekend crowds. He was dressed in slacks and a summery short-sleeved shirt – very casual and relaxed and certainly blending in well against the throng that carried him along. He had not noticed me so I slipped in behind him and followed in the hope that he might lead me to the revelation of some clandestine purpose. Sadly, whatever the reason for 'Steve's' presence had already occurred. He was headed back to the large car parking area and his civilian car. I made myself known to him at that point and can vividly remember the expression on his face – he was as red as a Tudor rose and obviously more shaken than a Martini as he tried to rationalise my presence. Had I seemingly popped up out of the blue, or was his worst-case scenario realised and my presence indicated something more sinister and potentially damaging? It was almost comical; it was certainly not the icy reaction of the spy of popular fiction. It was to be cherished as a unique scenario but was, of course, a complete coincidence. My conclusion was that he had definitely been engaged carrying out some form of operational tasking that, sadly, I had not been witness to, perhaps by only a matter of minutes – it must have been a relatively close call. At least I had spotted him and given him serious cause to reflect on the how and why. I wondered whether he had declared our encounter back in his office that Monday morning – I suspect not. Too many questions and too few answers. Not the best scenario for a Soviet intelligence officer to drop himself into without a better reason or explanation than this encounter generated.

I had already heard from another source that Maltsev had popped up in Moscow. This was not perhaps surprising in its own right, but that he had deliberately and pointedly made contact with an ex-USMLM tourer subsequently posted to the attaché staff at the US embassy was noteworthy. I had had no further contact, or even sighting, of 'Steve' when, a couple of years later, Martin contacted me.

He had been posted, after BRIXMIS, to the Joint Arms Control Implementation Group (JACIG), which had the responsibility of making snap inspections on former Soviet bases to ensure they were compliant with the 'Vienna Document' and the Treaty on Conventional Armed Forces in Europe. He had been part of a team conducting an inspection somewhere in the former USSR when whom should he meet, again out of that same blue, but Maltsev. Our friend had just seemingly literally popped up from nowhere. You can certainly make the comment that it is a small world if nothing else. It appears Martin's presence was not necessarily 'Steve's' motive for dropping in on the inspection as the two did not engage in any protracted conversation, but who really knows?*

Just as a point of interest, I declared all my personal contacts with suspected Soviet intelligence officers and provided regular, comprehensive, detailed updates to our friends. Tasking was occasionally pushed my way, including on one interesting occasion that I will come to.

I was invited to a little get-together organised in the headquarters. I had bought a couple of books written by a former Soviet GRU officer under the alias of Viktor Suvorov. His real name, well established in the public domain, was Vladimir Rezun. He had defected to British intelligence in 1978 – coincidentally the year I had joined the Army – from Geneva, where he had been a member of the GRU residency within the Permanent Mission of the USSR at the UN HQ in the city. He had been encouraged to write his autobiographical books to embarrass the Soviet Union, and as a conduit to release certain information into the public domain. Amongst the subjects he reveals, Rezun confirms the GRU regarded their three Military Missions based in West Germany as full-blown residencies supplementing the other

* It's clear to me now that Maltsev was a KGB officer, closing my open mind at the time, and in his time, more than likely, the KGB rep within the ranks of SERB. In his book *Mission: A Cold War Remembrance*, Thomas Wyckoff describes a chance sighting of him, once he (Maltsev) had left SERB, outside the Soviet barracks on Friedrich Ebert Strasse, known to the Missions as Potsdam 283, waiting to catch a bus. 'Frank', during one of our debriefs, identified this barracks as the local HQ for the KGB detachment based in Potsdam (see Annex A). Joining the dots, I may be miscalculating two plus two, but I rarely give too much credence to coincidences, at least in a professional context.

two in the Federal Republic of Germany, located in the Bonn embassy and the Cologne consulate. This revelation significantly elevated the Missions' relative importance and status and testified to their operational importance.* It had been arranged for Suvorov to visit the Berlin HQ and address a small hand-picked audience about his short career as an active intelligence officer. At the time, this was a rare and exciting opportunity, as defectors themselves were a relatively scarce commodity and generally were well protected and discreetly guarded assets.†

He was a short, squat individual with a strong Russian accent, and although it was an interesting precedent to meet face to face with a member of his species, I couldn't help feeling a slight sense of anticlimax and disappointment at the lack of hard bite his story packed. To be fair and balanced, that was because his career had been a very short one before he had been resettled. It did graphically, however, illustrate the sliding 'currency exchange rate' of the intelligence world – nearly a dozen years previously, a young, inexperienced major in the GRU was gold-plated. The time was now nearly upon us when, unless you were a high-ranking military officer or at least a more senior-level intelligence officer, even if the deal was buy one get one free, it was likely the offer of defecting would be very carefully considered and probably ultimately rejected. The days of the big fish would be in, whilst the small fry were either thrown back in or left to metaphorically die on the deck of the boat.

A strange event that I have often in idle moments thought about subsequently occurred during the interregnum that existed after the last Mission had re-crossed the Bridge, but not very long before the Missions disappeared from the Allied orbat.

I don't remember the occasion but the Chief, Brigadier Tim, and Steve and I were heading to Potsdam. We were in civilian clothes and in two vehicles, Steve accompanying the Chief, I think in his capacity as the duty Russian speaker. We had just crossed over at Glienicke and pulled into the side of the road on the main drag, the Route One, leading into the city, before we were due to turn off to the Mission House. A UAZ-469

---

* Debriefs of Oleg Gordievsky, a KGB officer who defected to the UK in 1985, confirmed the Soviet Missions' importance.

† Rezun has published a number of books but specifically two about military intelligence under his alias: *Inside Soviet Military Intelligence* in 1984 and *Aquarium* in 1985. His first work came out in 1981, *Liberators: My Life in the Soviet Army*, followed by *Inside the Soviet Army* in 1982.

was already waiting at the roadside. Inside, I am pretty sure, were both Polozov and Pliev from SERB, both excellent English speakers. The Chief exited his vehicle, leaving Steve alone, and climbed into the Russian jeep. He was there for about twenty minutes. I never did find out the topic of what must have been an in-depth and intense conversation, nor was any explanation given. It appeared, by the strange and unorthodox choreography, to have been planned and expected but what Steve and my roles had been – if indeed we performed any – mystified me. What were we deaf and mute witnesses to? I can only conclude that something noteworthy and untoward had occurred that remained very tightly discreet. I have a recollection that shortly afterwards, I think the very next day, the Chief was summoned to the WGF HQ at Zossen. Whether he received an interview without coffee is unclear but his tour officer interpreter that day had obviously been briefed to restrict any comments on the details of the meeting to a firm and enduring 'No comment'.

I have debated the issue with myself and wondered if Brigadier Tim had over-played something on the HUMINT front that he had not briefed me on, and it had blown up in his face. Knowing him and his enthusiasm for the cloak and dagger, I think this highly feasible. With the end at hand and a diplomatic démarche of little consequence, it is highly conceivable that he went all-out for glory and regrettably failed. Hence the unusual summons to the Soviet command headquarters with its potential spanking and the tight-lipped manner in which this occurrence was subsequently handled. I think I have the scenario straight in my head and can guess at the identity of the players around the table. For their part, the two Soviet 'Ps' were either warning the Chief or, more likely, explaining the error of his ways to him and passing on the invitation for what was to happen the following day. If I am right, I applaud Brigadier Tim's attitude and conduct. All very credible, and I think believable, but ultimately nothing more than idle speculation.

Whether any of these contacts, either on our side or from the Soviet perspective, led to any further exploitation, I have no knowledge, nor would I comment if I had any, or any additional understanding.

What I can state categorically, slightly tongue in cheek, is that at no point in my career, or subsequently, has any approach been made to me by any intelligence organisation, friend, foe, former foe or enemy masquerading as a friend. There was an oblique feeler put out to me a few years ago, long after I had left the Army, by an academic institution in Israel suggesting some consultancy partnership but I let that slide without any follow-up. I doubt very much if Israeli intelligence had even the tiniest interest in someone with my background and level of access then, but I decided to give a wide berth to any situation that could have created potential issues. This whole lack of any meaningful contact either attests to an appreciation of my incorruptibility; an estimation of my loyalty to the Crown, my country and the capitalist system; or the dispiriting realisation that my recruitment would yield little of value to anyone. I have often speculated and idly considered though, what, if anything, would have incited me to treason? I have always adhered to the clichéd belief that everyone has their 'price', but in intelligence terms the price always has an upper limit based on the potential value or, in other words, the assessment of access, current or potential, of an agent and the intelligence they can collect. Even intelligence agencies, and perversely those of the former Communist Bloc, were, and are, regulated by the laws of supply and demand and a perception of value for money, subjective though such judgements are. I know introspectively that my value would fall well below the size of any 'suitcase' of dollars that would induce me to consider betraying the values I hold dear. A better target for recruitment, rather than an operational military intelligence officer, would be a staff officer working in a large HQ – this is exactly where Soviet intelligence enjoyed success. Such targets would have better access to high-value intelligence and might more readily climb the pole to higher office, influence and access. So ultimately, a moot and distracting, but still intriguing, point.

15

Real Research

Deploying the footrest and pushing back into the comfort of my new www.armchairgeneral.co.uk recliner, with the benefit of hindsight that 1,000km and over thirty years bring, I feel it is necessary to make a bold statement. Regrettably, BRIXMIS did not deliver in terms of the intelligence it derived from human sources. JIS did more but, measured against the expectation that there is no limit to success, it too underperformed. As a Russian armchair general, I would not be drawing the same conclusion weighing up the contribution made by the Soviet Military Missions – more likely, I suspect, I would be proudly applauding their professional panache, courage and achievement. In the case of the British Mission, given its unique position and access, it absolutely could and should have done better. JIS worked hard and with focus, in its time, but from my perspective it disappointed too, unable to overcome the severe operational restrictions that circumstances imposed upon it.

That the Mission did not deliver was the consequence of a number of factors. The first relates to the tasking the Mission received, and to the culture this fostered, encouraged and perpetuated within the unit. In direct comparison to the Soviet Missions, who were consistently assessed as having a clear and active mandate, which they relentlessly prosecuted, the British Mission was not in any way directed by an officially sanctioned and established agent-related operational role. I would assert that, seemingly unlike the Soviets, neither our military nor our civilian leaders chose to exploit the opportunities the Mission's unique role generated. The freedom of access to both potential military and civilian human targets was either not appreciated or, more likely, ignored. Tasking remained restricted to operational and technical intelligence collection, and to liaison, even though, in the conduct of executing these liaison tasks, BRIXMIS personnel maintained unrivalled contact with Soviet officers. As I readily witnessed, civilian agencies had nowhere near the potential operational freedom that we enjoyed. And I would be certain the situation I observed had been no different from the constraints their 'forefathers' potentially chaffed against in the 1970s, '60s or '50s. The enduring reality was that the former DDR was an extremely difficult and restrictive environment for Allied civilian intelligence officers to operate within. Unlike the countries of the West, the closed nature of the DDR prevented free and easy movement. The only representatives of the West who enjoyed any degree of regular, routine and systematic access were the Military Missions: access guaranteed and protected by the post-war agreements.

For the duration of its life, the Mission remained unaffiliated to any intelligence agency. It certainly was not a maverick, but it operated independently, reporting to the various military staffs who were not themselves, generally, staffed by intelligence experts. It was they who projected and sustained a more narrow perception of the art of intelligence onto the Mission. This perpetuated the culture and discipline of the single, narrow focus on operational military intelligence collection. Peripherally, however, if and when any exploratory discussions might have taken place regarding a broader collection mandate, there may well have been push back from at least one national intelligence agency, less eager for the Mission to encroach on its hallowed agent-running turf, even where it might have acknowledged difficulties regarding its own success.

Further, and crucially, a major additional factor at play was the British approach to manning BRIXMIS that persisted from its inception to its disestablishment. Unlike the three Soviet Missions that were fully professional military intelligence units and, according to defector reporting, regarded as *rezidenturas* of the GRU and staffed accordingly, the British Mission, as I've already stated, was manned in line with the ethos that the amateur ruled. Put bluntly, and again boldly, the BRIXMIS establishment was less than fit for purpose. What I mean by this is that the blend of manpower skills was, in my opinion, inadequately balanced. BRIXMIS should have possessed a hard, rigid backbone of professional intelligence officers. Intelligence officers with both a capital and a small 'i'. Some of these professionals, those with a capital 'I', from the Intelligence Corps, should have had an agent-running background. There were enough post-war conflicts, even before the advent of Northern Ireland, where they could have received their operational schooling: Palestine, Malaya, Cyprus, Kenya, Aden. Latterly, in the last ten or fifteen years of the Mission's life, officers with a surveillance background from Northern Ireland should have augmented them, representing the small 'i'. The added value of their posting into the Mission would have been their bespoke 'recce' skills honed on the streets of Belfast or Londonderry, or the narrow lanes of County Tyrone – the Mission needed more than the odd one or two officers and SNCOs who served in BRIXMIS on a seemingly *ad hoc* and random basis. Reconnaissance, the Mission's bread and butter, is after all an established and NATO-acknowledged sub-set of human intelligence. Where were the HUMINT specialists?

In this revised manning model, the Int Corps agent handlers, I would advocate, should have constituted the bulk of the Mission's linguists, those with the balanced mix of Russian and German. They would have been at the forefront of the Mission's liaison role with SERB and the GSFG/WGF command, as well as an integral and significant part of the touring effort out on the ground over the Wall. Regardless of any wider agent handling-related role that may or may not have evolved, they would have had the professional aptitude to maximise the potential gain even from the routine liaison contacts with the Soviets, encouraging and steering the flow of HUMINT-related intelligence, adequately and effectively, routinely regulating these relationships. Making a key point regarding HUMINT human source operations: more specific and tailored operations against targets whose language one does not speak fluently just do not work if a handler attempting to build rapport and push a target towards meaningful HUMINT reporting cannot do so in the language of their contact, or in a shared common language such as English. Regardless of how good or how empathic an interpreter is, it is not possible to create the bond that cements agent and handler together. I can personally attest to this from my own experience working in the Mission, in the Balkans, in Sub-Saharan Africa and in the Middle East.

This hard, sharp edge should have then been augmented by the permanent presence of Special Forces personnel with their honed, broader reconnaissance skills set. Positively, this did seem to be a small and enduring feature of the BRIXMIS orbat, but we are talking about a very small percentage of touring personnel.

Despite any move towards specialisation, I want to make the point clearly that it would have been important to maintain a balance, thereby retaining a workable proportion of tourers, both officers and SNCOs, from the teeth arms in order to exploit their specialist conventional military knowledge and experience. As infantry, armour, artillery or engineer specialists, they would have brought with them the contextual professional knowledge that added value to the Mission's ground touring operations. In parallel, on the air side, tourers would, unquestionably, have been required to remain exclusively RAF non-HUMINT specialists – evaluation of Soviet aerial doctrine and flying skills required professional perspectives that only aircrew could bring. I would not have advocated any re-balancing or change here.

Given the structuring of tour pass allocation, a degree of numerical juggling would have been required that would have led to some sort of split between officers and SNCOs against my functional recommendations above. This would, however, only have constituted minor window dressing as an administrative exercise. And any establishment is always open to readjustment and fine-tuning if it is deemed to be necessary over time, either proactively or reactively.

With a cadre within the Mission who were experienced agent handlers, even if there had been no more aggressive accompanying doctrinal change in operational tasking focus, a dynamic would, without any doubt, have been created that would have produced pressure to change, or rather promote, a more proactive agent-related policy, even if this was restricted only to providing limited support to other agencies' operations. These 'experts' would have agitated and pushed upwards within the chain of command to obtain sanction for enhanced agent-related operations, even if the system remained ambivalent to this source of intelligence. Military officers are as skilled at lobbying their superiors as any other professional body. I suspect that at some point there might have been a convergence of ideas, those pushed upwards from the grass roots, and recognition of the validity of the approach from an officer senior enough in the hierarchy who would have acted as their sponsor and professional 'uncle'. Undoubtedly, a strategy and *modus operandi* would have evolved with, no doubt, operational ebbs and flows along the way, but playing a part in the production of a steady flow of human source intelligence. This, however, did not happen. But this is the food that nourishes the armchair general and brings him cheer as he pontificates from the comfort of his recliner.

If things had been different, what sort of intelligence could have been collected? Even informal relationships with Soviet personnel regardless of any attempts to formally recruit agents, could potentially have produced advance warning of issues or crises, or produced updates on events that were already unfolding. Such intelligence could have been valuable, informing the policy makers, especially during times of crisis and tension. The Mission would have been in a position to augment the intelligence generated by its *de visu* sightings with human source intelligence across a wide area of reporting. During the course of its operational life this could have included comment on the major international potential flashpoints, for example historically, regarding Soviet military reaction to the crises in Berlin in the 1950s, Czechoslovakia, the war in Afghanistan et al. No method of intelligence collection provides better access into the thought processes, objectives and plans of an adversary. At the tactical level, structured contact with Soviet sources could have provided more routine, but informed, input on tactical doctrine, new equipment, the state of morale, organisational changes etc., etc. Even broader generic-type intelligence like this would have played an important role in our better understanding of the threat posed by, and intention of, our enemy.

Establishing systematic contact with East German citizens across the spectrum of political, industrial or social activity would have produced the kind of diverse and varied background intelligence, by no means restricted to military topics, of interest to our political as well as our military leaders, just as occurred in the early years when BRIXMIS passed on intelligence relating to East German steel production. The Mission could have constructed a broad network of informants right across DDR society, from teachers to postal workers to farmers to ordinary workers. These relationships would have been valuable even if, in reality, they were superficial – from any dialogue, information can be drawn that can be processed into intelligence. More punchily, a more systematic targeted network of civilian sources working for the railways – rather than the piecemeal success that we had – who would have routinely reported on military train schedules, and those living close to barracks, training areas or key ports could have monitored military movements. Even the passage of low-level background information would have played a positive role in filling in those much-referenced gaps in that big intelligence jigsaw puzzle that continues to challenge our intelligence agencies.

Recruiting and running fully fledged agents would have required extremely careful operational control and been predicated against a very rigorous risk-versus-gain matrix. Perhaps this would have always been, in real terms, a step too far in terms of the risk appetite of UK Plc and our allies, but this decision should have been preceded by the degree of informed and rigorous debate that I believe did not take place at any stage during the Missions' lives.*

As it was, for most of its life, the Mission hierarchy avoided a collection discipline they neither understood nor probably trusted or even respected – 'spying', that rather unseemly, unbecoming side of military life. No doubt they were fearful of compromises they felt could have threatened the continued existence of the Missions, so restricted and muzzled any debate. But realistically, all collection operations bear inherent risks, require caution and the application of professional tactics and tradecraft – agent-related operations would not have engendered any new or novel imperatives. Frankly, I believe the failure to exploit BRIXMIS's full potential was based on a naïve and weak viewpoint, failing as it did, ultimately, to vector in the principle of reciprocity. The Soviets would not have risked creating a major issue over a tactic that they themselves are assessed as having so vigorously exploited and valued, wherever they operated – they, arguably, had more to lose. And anyway, we know they assumed we were actively engaged in these type of operations right up to the end.

In the event, a very small and unique window of opportunity emerged and was grasped to push forward agent-focussed HUMINT operations. I hope without sounding arrogant, my arrival in the Mission, as the only professional agent handler, coincided with the changes that the toppling of the Wall created. The Chief appreciated the new environment and the opportunities it presented and sanctioned a more aggressive approach, but time was short. And there was only so much one man could achieve personally or push and steer other like-minded, but fundamentally untrained, tourers

* An agent, in intelligence terminology, is a human asset that is formally recruited and under full control, run by an agent handler from an intelligence organisation. As such, agents are taskable, steerable and are rewarded for the intelligence they provide.

to undertake. As a part-time tourer, my opportunity for crossing into the new east after 9 November remained limited. But this was undoubtedly a period of tremendous potential. We faced significant opportunity in terms of point of contact: the potential to get alongside targets and build those nascent relationships from which fruit could mature. It was the time to exploit the growing potential motivating factors of unease and concern about the future that we knew, from our human contacts, were beginning to grow and fester within the ranks of WGF. And it was still the time when the mid-ranking officers and men we would be meeting were still considered viable and attractive targets. This reality quickly changed as 'devaluation' kicked in.

Paradoxically, once the Mission had been disestablished and JIS had assumed the operational role, I, for one, had complete freedom of access into the former DDR but our *modus operandi* and the change in our status meant contact with uniformed potential Russian military targets was extremely difficult to engineer, and certainly no longer routine as it had been for us as members of an officially accredited quasi-diplomatic Mission. JIS was a covert unit and, as such, should have been, and remained, invisible to the Soviets.

We pushed and pulled as best we could in those twilight years as the end approached, but the conclusion has to be that we also failed to produce any human source intelligence of any significant strategic value. At the tactical level of reporting, however, as the 'rump' in the final months of BRIXMIS, we were successfully adding to our targeting knowledge of a number of Soviet officers, including, of course, SERB, and we were able to get a very good look inside the Soviet and MfS anti-Mission operations, but there was so much more potential that we passed over. Where the Mission had achieved so much on the IMINT side with the quality of its technical collection, its HUMINT record was abjectly disappointing. Had BRIXMIS been designated and subordinated as a defence intelligence unit, the record surely would have been more substantive and significant back across the years, not just during the little sliver of time at the end. Concluding my assessment, despite my hopes and optimism, I am forced to tar JIS with the same brush. It too failed to deliver the scale of substantive human source intelligence that I initially felt was feasible. However, we can be thankful as professionals that at least we had this limited opportunity to work in so special an environment.

With the Wall down and its toppled mass getting progressively smaller and smaller as shards and whole chunks were malleted and sledge-hammered off, there was the acceptance at the top of the Mission that we could and should exploit the HUMINT potential that began to appear that, in the days of the old regime, we would have steered well clear of.* This positive, proactive view was not universally shared amongst the 'head-sheds', but Brigadier Tim had seen it all in the Province and, as I knew from my personal contacts with him there, he was well aware of the potential for scoring significant intelligence and had certainly supported me in South Det FRU when he was able.

Martin had done great work establishing and developing an excellent contact with the railway worker who ran a signals box that sat nicely on the strategically important rail line between Berlin and Leipzig. Great though this was, it was an opportunity target and not the consequence of deliberate systematic targeting. Martin had seized the opportunity that he had spotted. She lived in a small cottage next to the signals box. There had always been those brave individuals in the DDR who had helped Mission tours in some way where they were able, whether it was the farmer revving the engine of his decrepit tractor to pull a tour vehicle out of sucking mud or a householder signalling through their front window that the Sovs 'went that way'. We now, however, very quickly started to experience the reality, on a grander and wider scale, that with the Wall down, those who had shared anti-regime sentiments of any shade or hue, or who were at least sufficiently realistic enough to appreciate that the old attitudes had to, and would, die, started to help in any way they could. At the other extreme were those zealots who had been a dedicated part of the structure but now realised their world was done and sought to demonstrate their enthusiasm for the new order as a way to mitigate their pasts, and those, of course, who would rather be smothered than aid the forces of imperialism.

Whatever her motivation, this particular *Reichsbahn* employee chose to provide Martin with the timetables of military trains involved in the process of withdrawing assets from the country and returning them home

---

* This said, I have heard anecdotal tales that suggest in the early years of the Mission at least one or two of the more 'grounded' tour officers did run a number of unofficial contacts. Whether these were predominantly intelligence contacts or players on other fields is not clear to me. These were murky times and certainly there would have been a less-developed code of what did and did not constitute 'proper' or professional behaviour.

eastwards to Russia. This was absolutely superb intelligence, on two counts. It now meant BRIXMIS tours could be surgically tasked to occupy key rail vantage positions, unobtrusively shrouded in shadow, from which they could make key observations. And now, with this quality of reporting, we were consistently in the right place at the right time without the speculative time-consuming wait of pre-*Wende* days.

She would surreptitiously hand Martin the schedules when the latter pulled up in his G-Wagen at the signal box. Initially, Martin felt exposed conducting so bold and overt a contact and discussed with me the prospect of establishing a dead letter box, the DLB of classic espionage tradecraft. This would have negated face-to-face contact and involved the signaller placing the documents in a secluded hiding place for Martin to later empty. I was against this and ran the idea past Hamish Norrie, the chief G2 in Berlin at the time, the senior professional intelligence officer advising the commander of the Berlin garrison, for a second opinion. He too thought the risks of this approach outweighed maintaining direct contact – neither Martin nor the contact were trained in the use of DLBs, and the use of one was potentially much less secure and more dangerous, with the *Stasi* still operational, than conventional contact.[†]

Martin met her a few more times and continued to bring back invaluable intelligence, which was of course shared with the other two Missions. As recompense, and to maintain motivation, we organised for the contact and her family to come up to West Berlin for a social rapport-building meeting to cement the relationship. Martin and I met them and spent the day at Berlin Zoo, having lunch and generally a good time. It was a useful exercise. I presume that post-unification new technology employed by western *Deutsche Bahn* has replaced our former comrade and she may well be one of the substantial number of systemically unemployed in the eastern half of the country who now lament the passing of their Wall. If so, that is poor reward for her efforts.

It was a Sunday. I got a message to contact the Chief at home. I went round to his residence to meet him. As a couple, he was married to a super Australian woman, a former Qantas flight attendant. Both she and her

† FRU operators were trained in a host of these classic techniques, including brush contacts, but very, very rarely had the operational recourse to deploy them. The consideration and selection of tradecraft techniques was always a matter of what best got the job done in the most secure and effective manner – horses for courses.

best friend, with whom she regularly flew, married British Army officers. Brigadier Tim and his wife were warm and welcoming hosts even on *ad hoc* drop-ins such as this. There had been a walk-in, someone with unspecified links to the MfS offering us intelligence, the Chief briefed me, and I was to engineer a reverse contact, arrange a meet to assess the potential of the case and debrief him, squeezing every gram from this unexpected gift. This was exciting news, tantalising even against this sparse overview.

The reality of operational life in the city and the surrounding country was schizophrenic and rapidly changing, but there was still real work to be done. It looked increasingly that Germany would be reunited, the Soviets would complete a 100 per cent withdrawal of forces – even though the steady move of forces eastwards had begun, new equipment was still paradoxically being shipped and 'railed' into the old DDR. However, whilst this happened, there were still thousands of Russian troops and their equipment in-country; they were still a hard target defending and protecting their security, and, although in its death throes, there was still an MfS to counter Mission collection efforts even if their 'counter-revolutionary' stranglehold on the internal situation had no doubt loosened.

The walk-in turned out to be an officer in the MfS based in Potsdam. He had made contact with a liaison office in the Berlin garrison and then returned home after leaving a telephone contact number.

The next day I sat down and put together an appreciation for the Chief, presenting a number of options for meeting the contact clandestinely, citing the advantages and disadvantages of each – the classic military approach to tackling a situation. From the three options I drew up, I recommended the one that I considered to make the most sense operationally and in terms of good security. It involved a plan to securely meet the contact on the outskirts of Berlin in a quieter suburb he could easily reach, and in which I could factor in a counter-surveillance phase to ensure he was not being followed. The risks of meeting him seemed negligible and the prospect of some sort of PR set-up or provocation even less so, but it was professional to be professional.

Brigadier Jackson approved my plan and I made contact, calling on the number our walk-in had left.

That week the target came across from Potsdam as I had instructed him, walked the route I had briefed him to walk and passed through the three choke points where I was waiting to observe if he appeared to be being tailed. I recognised him based on the description he passed me over the phone and by the shopping bag I had instructed him to carry in his left hand. Our contact was by himself.

I met him on three other occasions, twice at his home on the outskirts of Potsdam with his wife, and we went out for dinner in the town on another meet. His flat was the typical small East German pre-fabricated apartment, frugally furnished with functional, utilitarian DDR chic, but spotlessly clean. On the dining table were the standard jarred pickled vegetables and tinned meat. I was made a welcome guest, the first Westerner to set foot in his home.

'Frank' was a young MfS officer and described his training and the organisation and mission of the *Stasi* in the Potsdam *Bezirk*, or region. Although still formally a member of the Ministry, he was no longer actively employed. Physically, he was 30, sported short brown hair and wore metal-framed glasses. He was completely unassuming looking. He came across as a rather melancholic character but fundamentally open and honest. Mentally, I concluded that since the demise of the regime, he had had time to reflect on issues of good versus evil, the nature of the former SED leadership and the role and record of his employer, no doubt substantially influenced by the Western media that he now would have had even easier and routine access to. He struck me as being influenced by feelings of personal guilt for his association with the *Stasi* and for the impact it had had on 'the lives of others', even if they had been critics of the state.* I thought he was genuine and not motivated by any attempt to 'suck' up to the Allies and disassociate himself from any past collective or personal responsibility.

I enjoyed our meetings, learned some invaluable and fascinating insights into life in the east and particularly how the *Stasi* functioned, especially with relation to the Missions, and wrote up a number of reports, including a consolidated overview drawing together the major points of intelligence for the Chief to disseminate as he saw fit.

A few years later I had reached the end of Tony Geraghty's book on the Mission and was reading one of the annexes. It was a report of the debriefing of a MfS officer. As I read, I thought it was pretty good, then

---

* Directed by the brilliant Florian von Donnersmarck, *The Lives of Others* is a powerful and haunting depiction of the *Stasi* at work. Well worth watching in the original German with subtitles.

began experiencing a strange sense of déjà vu. Sentences began to feel familiar. It finally clicked that Geraghty's annex was my consolidated report. I was a trifle surprised, then began to wonder how he had got hold of what was then a classified document. Obviously in the feeding frenzy of the post-Mission exposé, as the official history was put down on paper and the archives thrown open, the cats, chickens and horses were allowed, if not encouraged, to vacate the bag, coop and stable. I have reclaimed my work and have stuck it in an annex at the end of this – see Annex A.

'Frank' was able to outline how the *Stasi* was organised in Potsdam and how its eighty-six intelligence operatives were deployed in their attack on the Missions, their main target. He highlighted the three main approaches: monitoring incoming and outgoing calls from the Mission Houses, mobile surveillance of tours and reporting from 'unofficial members' – the MfS euphemism for agents, working as shopkeepers, petrol station attendants and ordinary citizens living adjacent to locations regarded as being potential Mission targets. 'Frank' strongly reiterated the point made at the strategic level, mirrored at the tactical level in Potsdam, that with regard to the MfS, as an organisation, it collected so much raw information that it was unable to process it efficiently and struggled woefully to turn it into exploitable intelligence.

He identified that his anti-Mission unit worked under *Abteilung* 5 of *Hauptabteilung* VIII in MfS HQ in East Berlin. *Abteilung* 5 was further organised into seven separate *Referats*, including the agent handlers; two regional surveillance departments (north and south), plus a team controlling the surveillance callsigns on the ground by radio; the eavesdropping unit and its subordinate analysis cell; a team that interfaced with the civilian population increasing awareness of Mission activity and an admin section.

Stasi anti-Mission activities aimed to, first and foremost, prevent or disrupt our attempts to collect intelligence. This explained the dogged tactics of the surveillance teams. By gluing themselves to our rear bumpers, they sought to intimidate us and make it so high risk to attack and photograph a target that we would abort our efforts. Obviously, this did not happen, and tours brought back enormous quantities of raw intelligence, but it did impinge on the way we did business tactically. The analysis cell would study touring patterns in general, and those of individual tour officers in detail, in order to attempt to task the surveillance teams more surgically and proactively, ultimately to initiate planning to orchestrate the planned, coordinated detention of a tour. On occasion, the Soviets passed on advance notification to Potsdam that new equipment would be deploying to the DDR so that the surveillance teams could work to protect it – and frustrate our collection efforts. This often required team personnel to deploy into the field for weeks at a time, living out of caravans or tents.

If they failed to deter Mission collection operations, the aim was to catch tours in flagrante and obtain photographic proof of their espionage activities, which could then be passed to the Soviets as the catalyst for protests that could be raised against individual tourers via SERB.

The MfS appeared to be fixated by its assessment that the Missions were closely tied into their respective national intelligence agencies and were conducting agent handling operations on their behalf. These links it sought to establish and prove. A high-priority objective was therefore to mount deliberate, planned, aggressive detentions that would enable documents and equipment, carried in Mission tour vehicles, to be snatched, detailing these suspected links and affiliations. This variety of planned, orchestrated detention was completely independent from the routine opportunity detentions mounted *ad hoc* by Soviet forces on the ground. Planned detentions might also be considered in retaliation as reciprocity for the detention of Soviet Mission teams in West Germany. 'Frank' revealed a special squad of GRU operatives and/or special forces (*Spetznaz*) had been established, based in the Soviet headquarters in Zossen Wünsdorf, to mount these operations. The squad was able to co-opt local MfS personnel if they needed to surge manpower. They would deploy in Soviet uniforms. Interestingly, the Soviet high command retained the option of detaining tours, even after the demise of the MfS in February 1990.

The final Mission-related aim of state security was to gather information in order to build profiles of individual tour officers and identify their specialist skills and areas of interest. The *Stasi* paid no heed to uniform or any other insignia. This confirmed my scepticism, as I have already outlined, that the policy of disguising the Special Forces and professional intelligence personnel in BRIXMIS, by forcing them to adopt the cap badge of another corps or regiment, made no real sense. 'Frank' reported that these files were passed on to the foreign intelligence service (the HVA) and to the central analysis and information group (ZAIG) in MfS headquarters in East Berlin, controlled by the Deputy Minister for State Security. ZAIG, in turn, would liaise with friendly Warsaw Pact foreign intelligence services to aid the identification of former Mission personnel resurfacing in embassies or other staffs behind the Iron Curtain or in friendly countries in the future. No doubt SERB worked hard to augment these profiles in its liaison dealings with us.

He also passed on how the *Stasi* perceived Mission activity. They saw our prime role as the collection of military-related data, primarily against the Soviets. They had this right but over-estimated our relationships to the national agencies. They were also incorrect in assessing that tours had always intercepted air-to-ground communications to target aircraft, although this did change in 1986, greatly increasing the efficiency of tours during the last three years. They were also wrong in their belief that we deployed ground sensors to detect the movement of vehicles. Perhaps this assumption confirmed this as a Soviet Mission tactic – you see the world from where you stand. They were correct, however, in reporting Mission interest in economic targets, but wrongly assumed the Mission's non-existent long-standing role in running human sources included meeting agents directly out in the field and servicing dead letter boxes – again, no doubt using their own activity as a mirror through which to gauge the actions of others, and surely a conclusion predicated on their sure knowledge that the Soviets' western Missions engaged in active agent-support roles. In the context of intelligence parlance, this analytical approach is known as 'mirror-imaging'.

The breakdown 'Frank' provided of the eavesdropping operation was extremely interesting. Every region within the country had its own organic eavesdropping unit monitoring internal security. The section based in Potsdam, however, was exclusively focussed on anti-Mission tasks. 'Frank' worked in the monitoring section. Technical installation of bugging equipment was the responsibility of another unit. They directly tasked the post office engineers who maintained the six lines going into and coming out of the three Mission Houses, which were then diverted and extended to the monitoring section in Beyerstrasse in Potsdam where 'Frank' worked. The inductive technique used to monitor the lines meant there was no indication to line users that they were being monitored.

The monitoring operation was technically unsophisticated but reliable, using voice-activated cassette players to record conversations. These were transcribed directly by the two-man team on duty during working hours (source plus one), and a daily digest was disseminated at the end of each day. Calls intercepted during silent hours were transcribed the following day unless the duty officer assessed that the call(s) required immediate translation. In this event, he called one of the transcribers into the office for immediate action. The daily report was disseminated by the *Stasi* lieutenant colonel supervising the eavesdropping unit. Recipients were the anti-Mission unit commanding officer – a full colonel – the analysis cell in Potsdam, and the KGB detachment based in the outskirts of the city. Unidentified numbers ringing into the Mission Houses from the West were rung by 'unofficial members' to establish the identity of the callers. 'Frank' revealed that USMLM was the most insecure in terms of telephone security, while the French gave away the least amount of exploitable information.

With the dissolution of the *Stasi*, all files were handed over to the KGB.

As a footnote, we discussed the final days of the *Stasi* before its dissolution. 'Frank' stated that, initially, with the toppling of the Wall, there was some naïve optimism that the organisation would survive. But as political reform gathered pace, by December 1989 it was clear there was absolutely no future for it, or its members as state security operatives. By this date all MfS anti-Mission activity had ceased. In February everyone was sent home and, with the transfer of all paper records by lorry to the KGB, the front door at Beyerstrasse was locked for the last time.

The KGB subsequently approached a number of MfS operators in an attempt to recruit them, but by the end of the year, 1990, 'Frank' reported that even they were inactive with little to do. Significantly, he reported that the comment coming out of the KGB office was the desire for a military coup in Moscow to remove Gorbachev.

To supplement the MfS operative angle, I 'recruited' one of the staff working in the Mission House. We met a number of times just over the 'frontier' in the west of the city. This particular worker had worked for the Russians – others, I had learned, reported to the *Stasi*, although it was not clear to me how any demarcation between the KGB and MfS was decided upon. I assumed that information gained was pooled and shared, not that I believe it would have amounted to much of substance due to our institutionalised caution and defensive attitude with all things House-wise. They, the House staff, had absolutely no chance to refuse to cooperate with the intelligence services and I, for one, held absolutely nothing against them for what they did – a refusal to work for Big Brother would have meant no work anywhere else and a negative annotation in their personal file. Beyond these practical issues, it is important to maintain perspective – our

House staff were doing nothing more than serving *their* state and accepting tasking from one of *its* government agencies that was framed to preserve their security and do harm to us, their sworn enemies. Beware victor's justice – I have always sniffed cautiously in its general direction. Clearly, it carries with it two distinct faces, and the victors are often not that securely protected from future defeats.*

What I learned generally supported the take from 'Frank' and fluffed his debriefs out a bit at the edges.

One amusing story, concerning the source's handler, that made us both chuckle was the tale of one particular meeting in this Russian officer's office in Potsdam – I was not able to positively identify the officer but did rule out categorically any member of SERB by showing the source a photo montage of its members (one of the pictures amongst the selection of photographs). I think it is almost certain that the handler was a member of the KGB team based in the city, highlighted by 'Frank'. During the ensuing debrief in question, there was an audible 'click' as the concealed tape recorder ran out of tape and switched itself off. I could picture the scene, the momentary unavoidable lapse in the conversation, the instant colouring of cheeks, and sympathising with the handling officer's embarrassment. But he was guilty of either sloppy stage management or a general apathy regarding the importance of the case he was running.

As the clock ticked down on the continued life of BRIXMIS, two HUMINT-related stories remain firmly and vividly in my memory, but for diametrically opposed reasons: the first because of its professional impact and nature, the second because of the comic unprofessional farce from start to finish. In my mind, the latter wholeheartedly reinforces the premise that fast jets should only be flown by experienced steely pilots, minesweepers commanded by salty naval officers, infantry soldiers led in the attack by gung-ho infantry officers and HUMINT operations planned and executed by trained, professional intelligence operators.

I had had a call that a Soviet full colonel, an air force pilot, who, according to information – no further explanation – was 'of interest'. These were still the days when an officer of this mid-level rank excited significant

---

* Interestingly, at least at some time in the past, it appeared that all House employees were actually members of the *Volkspolizei*, the *VoPo*. It is unclear if this situation lapsed and, if so, when? But it was my assessment that certainly, latterly, House staff were 'civilians' not plain clothes policemen.

interest and was deemed of 'value'; this changed rapidly, as I've said, over the next months as the whole Eastern Bloc opened up like a rotting watermelon. I was briefed that the officer highlighted was based to the west of Berlin near the home to 3 Shock Army, BAOR's direct adversary, staring at them across the north German plain. There was little else to lock me on except the intriguing snippet that the officer in question owned a Husky-type sled dog, which was relatively unusual for a Soviet officer in WGF. I was requested to travel there and 'trail my coat' to see if the officer of interest, now designated the target, broke cover and made contact. I agreed that I would take the task on and tour down there in the next couple of days.

Chief BRIXMIS approved the plan and I asked Martin, as a fellow intelligence officer, Russian speaker and friend I could trust, to accompany me. I set to work on formulating a plan.

We set out in a G-Wagen to drive to the target location, formerly important for its proximity to the Letzlinger Heide training area, which had been, and was technically still, covered by a PRA. I had passed through the area, albeit not through the city centre, several times before on the British Military Train, another anachronism of the Cold War. The train ran daily to symbolically maintain access between the west and divided Berlin. Its passengers were restricted to British servicemen and their families. The restaurant on board was decent, and the whole trip was a very worthwhile experience as one meandered across the DDR – one of the great train journeys of the world, whoever hosted that old TV series. Sometimes train passengers witnessed significant troop-related sightings, which were always logged and reported by the military staff operating the train. On one of my trips we were just entering a section of dense forest and there, up on one of the forest hunting/observation-type platforms at rail-side, was an East German flasher merrily exposing himself, contentedly waving to the line of carriages – he was armed with the smallest-calibre weapon I had seen, to date, in the DDR. It was certainly not a scoop. I probably waved back, with my hand.

We planned to stay a couple of days so that our coat hem got well and truly dirty, and we optimised the chances of our man speculatively making contact, so checked into our typically socialist, modernistic and soulless hotel in the city centre.

We quickly made contact with the local *Kommandant* out of courtesy and as a potentially useful point of contact for future activity – this was always standard practice when practicable. He was a friendly, professional-looking infantry lieutenant colonel. In his office we explained we were in town on a goodwill fly-the-flag mission and would be there for a couple of nights, causing him no concerns or drama. He seemed to accept this at face value – if he did not, he certainly had the good manners not to throw his doubts in our faces.

That night we went out for dinner altogether in the city and had a very pleasant and open evening chatting about family, country and the current geopolitical situation. For fun, I had a brief drive of the *Kommandant*'s personal UAZ-469 jeep – not in actual fact a bad vehicle to be behind the wheel of.

The next day, Martin and I made ourselves conspicuous around town, including a visit to the Soviet military cemetery located on the edge of the city centre, a mournful but well-maintained site. The country was dotted with them and I usually made a point, if time allowed, to stop and pay my respects when I could. I have to admit that I was unduly affected by a headstone erected in memory of a 12-year-old girl, the daughter of a Soviet officer, who had died in the city in 1947, presumably accompanying her parents on her father's posting. It struck me as so sad and wasteful a fate to have survived the horrors of the Great Patriotic War, then to have been the victim of an accident or childhood illness, dying so far from home amid the even greater privation of post-war Germany.* Our presence must have been widely recorded, and maybe reported, but we were not approached by any Soviet officer, with dog or without, or anyone else for that matter.

We had agreed to meet up with the *Kommandant* again later that day. When we entered his office, as planned, he was accompanied by another officer, distinctive in his Soviet Air Force uniform, a wiry, fit-looking individual with slightly longer dirty-blond hair, open and cheery but at the same time displaying an underlying reserve, a distinct Asiatic cast to his eyes denoting his probable Siberian background. On his chest were pilot wings

* Military cemeteries obviously have more meaning to soldiers. I have always found them intensely moving places, especially the beautifully maintained row upon row of brilliant white headstones in France and Belgium. I was perhaps most personally affected by the small graveyard on the outskirts of Tirana's city centre. The graves there date from the Second World War and house the bodies of aircrew and irregular members of the British Army who fought in Albania against the Axis. I had just arrived, by myself, to cover the war in the north of the country and across the border in Kosovo, ultimately to link up with the UCK/KLA, reporting back to the Defence Intelligence Staff in Whitehall. It was bad psychology to stumble across the graves of these brave men in the circumstances. I was, however, able to report back to the War Graves Commission that the plot was being well cared for.

and an impressive collage of medal ribbons. We four sat down and chatted away happily, building on our acquaintanceship with the *Kommandant* and getting to know the new officer.

During the course of our chat, the newcomer challenged me to admit that I was a member of some special force. I asked him what made him think that and he replied that my personal carriage suggested some degree of special training. I was slightly taken aback as I have always slouched somewhat to minimise any military bearing. Anyway, I admitted I was an intelligence officer. In the circumstances, as we were looking to flush out the target, I figured such an acknowledgement could only be helpful.

Then suddenly, without preamble or introduction, as if awkwardly and clumsily scripted, the conversation turned on its end and our new friend began talking about his dog. A confessed dog lover, which I am not, when I asked him what breed his dog was and whether it was a big one or something more compact, he confirmed that his was a former sled dog that his brother had bred back home. I nearly fell off my seat with this thunderbolt from nowhere. I hope I managed to keep the stunned shock from my face. As I caught Martin's eye momentarily, it looked like, incredibly, we had located our man – 'British military intelligence does it again,' I smiled to myself. This was a real stroke of luck and, seemingly, an amazing coincidence, or had our presence encouraged him to initiate this face to face?

We chatted on, stressing to remain composed and normal. Eventually, after what seemed an age of restraint, I turned the conversation to the following day, our last full one in town, and gave the pair a brief overview of our itinerary. I outlined that we would be visiting the garrison cemetery in the morning from about 10 a.m., making it clear to our new friend, by implication, that he would be able to run into us in this quiet and discreet location if he so chose. Soon after, we took our leave, still unable to accept the coincidence and seeming good fortune of this chance meeting and keenly anticipating the morrow to see if we had indeed struck gold.

That evening we had a quiet dinner and went to have a look at a disco in the city – we were, after all, on the face of it, on a cultural tour. Bizarrely, at some point in the evening, dressed as we were in our green, lightweight BRIXMIS gear with our Union flag arm badges, I was asked if we were from the local fire brigade. We were in a sheltered location full of people who had lived very sheltered lives. But I had to laugh anyway. It was a quiet night but fun to see the inside of what was still effectively a DDR night out.

The next morning, we had breakfast and headed back to the cemetery. It was the perfect location for the brief meeting we planned to have, combining good security with a workable cover story for both parties if required. And the weather was dry and inviting – perfect. We strolled around attempting to appear calm and disinterested, inwardly as tense and taut as the strings in a grand piano. But the setting was peaceful and calming as we re-inspected the headstones and monuments. I was again drawn to the melancholy forlornness of the young girl's grave as we remained vigilant and poised to exploit the situation if our man made the appearance that surely he was driven to make. Whilst we waited, I couldn't block the same questions that kept turning in my mind. Was he our man, or was the mention of the dog the wildest coincidence? Would he come if he was? Or had he got cold feet during the night now that the moment was upon him, or at some point since he had become 'of interest'? As the clock ticked by, we were surely close to resolution one way or the other.

However, as midday tolled in a nearby church tower, it looked increasingly like we were to remain the only visitors that morning, wandering without purpose amongst the graves.

As if vainly attempting to squeeze opportunity out of the situation where none now no longer existed, I looked one last time at my watch. It was indeed now past our cut-off time. Reluctantly, I had one last searching sweep around and aborted the operation. He was not coming. We had jumped to conclusions, although not unreasonably, or else the intelligence had been faulty to begin with. We had no way of knowing. It was a tough let down and awoke those emotions common enough from FRU days when agents or targets failed to meet. I was happy, however, that we had seemingly successfully identified the target, but naturally frustrated that we had not been able to nail the second, and crucial, part of the mission. The chance to talk would really have covered the cost of the trip. On balance, with reflection, I think we had done well – the operation was well planned, well executed and no phase had led to any compromise. On top of those plusses, we had established a good relationship with the *Kommandant*, an important figure in the garrison, should we need to return at a future date. But there was no follow-up phase to discuss.

We made our way back to Berlin, and I briefed in the information. I heard nothing further on the matter.

The next tale I will keep brief and make no comment on the identity of those involved to protect the (not so) innocent. That said, I still have to smile to myself after all these years when I conjure up the images, around which this farce hangs, that still return to entertain me. I have met the main protagonist several times over the ensuing years, and each time we returned to these events, I have had to laugh and laugh and laugh. But the seriousness of the whole thing could have landed everyone in big trouble and conceivably shut things down, in terms of post-Mission activity, even before they had begun.

It was late winter and BRIXMIS continued to hang on to its name with its fingertips as the new year drew, inexorably, still closer yet. There were just days to go before the Mission furled its flags, put out the cat for the last time, finally closed its doors and vacated the field. The team in question, two Mission members, had done really well to date and had had a couple of meetings out on the ground with a Soviet officer, a lieutenant colonel. Had they not already proved themselves capable of following tasking and conducting meetings and debriefs, the pair most certainly would not have been deployed on so potentially sensitive a task. The Soviet officer, their quarry, was concerned about the future and seemed keen to talk more. We assessed that we were not facing the likelihood of any sort of provocation, so decided to meet again. The dynamic duo planned, prepped, briefed and deployed.

To cut a long story short, I got a call later that same evening of the day the pair were attending this third meet. It was the senior of the two, and he asked me to meet him in the secure but neutral environment of the warrant officer's and sergeants' mess in the Olympic Stadium complex as soon as possible. This sounded interesting but not unduly alarming as the pair were obviously back at base safe and sound and the call had not imparted any dramatic headline news. Nevertheless, I took a last mouthful, returned my dinner to the oven and set out straight away, only a six-minute drive at most. When I think back over my career, a business-related call or other matter like this would always take priority and preference over anything going on simultaneously in my private life. There was no compromise. I know I was never an easy person to live with.

The two of us made ourselves comfortable, sitting outside despite the cold to avoid any complications of the niceties of etiquette surrounding officers visiting the mess, and he proceeded to brief me on the day's meeting. As he began talking, taking me through the events still fresh and vibrant, I grew more and more incredulous as he continued his briefing.

They had met as planned at the pre-arranged RV point and all appeared to be fine and dandy – a good start. The Russian officer was happy to see them, and obviously the good degree of rapport that had been established was steering the relationship in a potentially positive direction. He produced a bottle of vodka and the three sat in his vehicle as they chatted amicably away, toasting and steadfastly working to reduce the distance and chill between hammer and sickle and lion and unicorn – an even better start and good progress.

At some point the pace stepped up a gear and the colonel suggested they return to his barracks to his accommodation to carry on the meeting in more relaxed, convivial and warmer surroundings – time for some quick thinking and assessment. The senior of our team agreed, no doubt conducting his rapid fire risk assessment, although obviously now in some state of alcohol-induced disrepair. If I remember correctly, they drove their Mission vehicle following the contact's UAZ. Progress now begins to flicker distinctly towards the red of the negative zone on the dial of the odometer. They obviously made it successfully inside through the Soviet main gate, parked up, no doubt crookedly as both had been drinking, and repaired to the Russian's mess-type room, what the Americans call 'bachelor officer quarters'. No doubt shot continued to follow shot and toast followed toast. We were now well and truly in potentially deep, dark and dangerous water. This would have been the time to man the life raft and push off back towards Berlin. But our sailors opted to remain on board as their craft began to take on more and more vodka.

At this stage in the story, time starts to telescope and detail becomes somewhat fuzzy around the edges but, incredibly, our hero, a senior non-commissioned officer in the British Army, decides, for whatever reason and with whatever logic, perhaps with some prompting and encouragement – maybe from our Russian friend, maybe from his teammate or maybe from a tag team of both – that it is a good idea to exchange uniforms with the contact. We were now under water, sinking fast, our lungs beginning to fill with icy water – we are starting to drown. I just cannot fathom, then or now, what pattern or sequence of electrical signals in their brains persuaded them that this was a sound, reasonable or acceptable course of action. It falls so far outside of logical, professional behaviour, even for non-specialist HUMINT intelligence officers. However, swap uniforms he does, a lucky coincidence that they share similar collar and inside leg measurements. He then, extrapolating the same logic, decides to leave the relative security of the mess room, and again perhaps with encouragement, succumbing to peer pressure, proceeds off towards the dining room or cookhouse of the barracks.

So we now had the wildly irregular, perhaps even unique, scenario of a British Army soldier, wearing the Red Army uniform of a Soviet lieutenant colonel, weaving drunkenly towards an official building in a Russian barracks in the heart of East Germany, with the intent of entering it and no doubt having something to eat.

A fair question to ask at this stage of the telling is, how did I know that this was not the fabrication or false memory of a mind distorted and corrupted by vodka? That is actually an easy one to answer. My hero had already surreptitiously had the incriminating photographs developed by Special Section, which clearly and unequivocally showed him in-frame and, perhaps surprisingly, in focus, in a number of shots taken from an upper-level window with enough accompanying background exhibits to confirm that, pre-Photoshop, we were inside a Soviet barracks and definitely cross-dressed.

In my complete and utter shock and stupefaction, I was totally unable to control my laughter at these absolutely incredible, surreal, unreal photos. If I hadn't laughed, surely I would have suffocated. I struggled to process, let alone believe, the story snapshotted on these oblongs of photographic paper. I could not believe that what had clearly happened had *happened*, nor what had possessed this normally calm, level-headed professional to act in so random, ill-informed and, frankly, so reckless a manner. It was just incredible. I remember sitting there, my head just shaking from side to side. It was a scene of complete and utter farce, the best comedy that Monty Python had ever scripted – why does everyone consistently reference the 'Parrot Sketch'? As a comedy scene, it had everything – utterly, utterly hilarious, utterly priceless – but as a military intelligence-gathering operation, chillingly scary and absolutely unacceptable. It is the sense of risk and daring, the potential revelation of the other flipside of a situation that enhances our nervous reaction as we laugh. The other side of this situation had had the potential to end with utter and complete disaster as our team was arrested and the Soviet brass in Zossen struggled to fire off a demand to the Chief quick enough, demanding an explanation. It was readily apparent to me that, in the collateral damage caused by the explosion, I would have likely found myself severely injured and that the shock waves would likely have rocked Hamish's boat too. As for the Chief, he would have been packing up his married quarter for pastures new even sooner than he expected.

Incredibly, this did not happen and our team managed to re-dress in the right clothes, in time, in the right order, the right way round and right side out, and to somehow leave the barracks without provoking an incident or accident of any kind and return to Berlin unscathed.

I sternly instructed that all photographic evidence of the escapade, in these pre-digital days of photographic negatives, be destroyed, and that he should never breathe a word of what had actually happened to another soul in Berlin. I said I would speak to his junior co-conspirator. I doubly hoped that no one in Special had taken the trouble to give the frames any more than a cursory passing glance and that no report would emerge from Russian sources to let the hounds off their leash.

In the event, the whole thing blew over, but thanks are due for so much mirth and humour, and for that eerie sensation as the chills raced up and down my back when I speculated on the way the whole escapade could so easily have ended. I still have not been treated to a better, more incredible story in any context, at anytime, anywhere in the world where I have served or since worked. Wonderful, wonderful images … Needless to say, neither of the twins ever sat again in a cockpit behind the controls of a fast jet, stood on the bridge of a minesweeper, fixed bayonets and led an infantry charge against the enemy or, more pointedly, deployed on any HUMINT operation that involved planned face-to-face meetings with Russian military personnel whilst they remained with the Mission or were carried over to serve with JIS.

As a final note on these HUMINT forays in the dying days of the Mission's existence, I had struck up a rapport with a previously unknown Russian colonel based in one of the garrisons in the local area. It seemed likely that he was an intelligence officer and I was keen to pursue the contact. The colonel seemed equally motivated to meet again, but at this stage his intent was not clear. We discussed possible venues for our next meeting. He suggested I cross over into the east and we meet in the officers' club in his Potsdam barracks, where we could enjoy the steam bath before some drinks and a relaxed chat. I thought about the idea for about one and a half seconds before concluding that it was not a plan that accorded with my concept of a sound rapport-building exercise, no matter what the motivations of this contact were. This was a scenario that could go horribly wrong very quickly. In the end we did not beat the clock and, unfortunately, did not meet again.

These were unique times and unique situations in classic HUMINT tradecraft terms. We felt the pressure that time was very much up against us and that we had a window of opportunity that would very soon begin to narrow, shift and ultimately close. We were eager to push and see what we could achieve in terms of intelligence gained and provide the bridge for any officer who considered defection or, better still, recruitment as an agent as the solution to the anxiety and concern that clearly existed as to the future in the new Russia. We knew we had the support we needed for this, should it be required, and I ensured we fed in regular updates on our activities.

In the event, the targets we were looking at very quickly disappeared off the radar scope as the system back in Russia imploded and secrets began leaking from all directions and sources from within the former Warsaw Pact. It was indeed a completely new intelligence playing field, subject to revised rules of the game and equipped with new goal posts. Whereas a lieutenant colonel had been rated as a desirable target, sights recalibrated very quickly, and if you were not a senior officer, military or intelligence service-wise, you were a nobody and not worth the operational candle. So, in the greater scheme of things, what we tried to achieve, both in terms of relationship building and the impact of intelligence reporting, very quickly lost any relevance as our time surely ran out anyway. But the point is that we tried nevertheless – for me, that is the most important thing regardless of success. We identified opportunity, we considered risk and we secured the necessary backing of the hierarchy. It could have ended differently and we all, collectively, could have been sitting down to tea with our medals. But that was not to be.

16

A Blot on the Landscape

One blot on the geopolitical landscape had looked set to potentially curtail my time in Berlin. This was the situation rapidly unravelling in the Middle East and the impending storm that continued to intensify with Iraq's invasion of Kuwait.

In November 1990 I had been briefed that it was likely I would deploy to Saudi Arabia, to Riyadh, to join the HQ there and perform a HUMINT coordination role in anticipation of a ground campaign to evict the invader. The job sounded interesting and the location a fascinating opportunity to get inside the Desert Kingdom, but I feared that my posting in Berlin would be over as soon as I stepped with one foot over the city limits. As the decision over the future of BRIXMIS was near resolution and the wider post-Mission future decided, I would be outside of the loop and soon forgotten. And I was confident in my analysis that war would not come and that Saddam Hussain would back down, logic and common sense triumphing, as he acquiesced to Allied pressure. I did not believe that he would risk the demise of his regime for the prospect of a mirage-like illusion; Iraq's claim to Kuwaiti territory was justified solely in the minds of the Iraqi leadership. Ultimately, my future operating in the former DDR would have been sacrificed for nothing. I was, of course, wrong in my assessment, and it was my confidence that shimmered without substance in the desert sand. Hussain over-assessed the importance of having been supported by the west in the Iran–Iraq War, underestimated the determination and resolve of Western leaders to preserve Kuwait's independence and sovereignty and perhaps lost the battle with the nihilistic streak running through his own psyche. I assessed in error that it would all be for nothing, but of course, as soldiers, we had to go where ordered to go.

My personal situation firmed up, but instead of being posted to Riyadh, and to a job where my background had some relevancy, I was informed I would now be parachuted in – not literally – to General Rupert Smith's headquarters as the SO3 G2, a standard, routine, low-level operational intelligence staff position. This made absolutely zero sense to me, and I was surprised and disappointed, in fact angry, at my own Corps' lack of vision. It must have been within the realms of a basic square-peg-to-square-hole philosophy to find someone more suited for an HQ role, especially set against the importance of filling an intelligence-gathering role. I had zero knowledge and experience of conventional war fighting, except what I had learned at Sandhurst and JDSC on my junior staff course, but I did know a bit about HUMINT ops. I did not know the difference between a divisional HQ and a McDonald's drive-thru, but BAOR was awash with bright, young operational intelligence staff officers sitting in the various brigade and divisional headquarters who would have excelled in this sandy role.

My posting order detailed my departure date as Boxing Day. I tried to look on the bright side. Surely I would not be marooned in the desert for more than a couple of weeks before normality returned, the *status quo ante bellum* restored, and Iraqi forces pulled back to Baghdad, or at least Basra – too short a period for me to be forgotten back in Berlin. I could then hopefully come back to real operations as the Soviets continued their withdrawal from the former east. Regardless, I began to consider the kit list that was appended to my joining instructions and what extra bits and pieces would make my time away in the sand more tolerable. I noted that a bucket and spade was not mandatory.

I had been having some trouble with my left arm that playing rugby appeared to be exacerbating. It was an issue with the tendon in my lower arm, about 8in above the wrist, and when active was painful enough to disturb my sleep. I thought I had better get to the doctor and pick up some painkillers and anti-inflammatories, or any more bespoke drug, just in case this would be problematical when I got to my new desert location. I made the necessary appointment and popped out of the office quickly to pick up my pills.

The MO, or medical officer as military doctors are labelled, was a Canadian civilian GP contracted to work for the Army Medical Services. I explained the issue and my rationale for making the appointment with him. As we talked, it became apparent he was ideologically opposed to the prospect of war with Iraq and looked at me, as I began to comprehend, as some sort of target for a one-man 'civil disobedience' campaign. Before I knew it, he had directed that, in order to tackle the issue with my tendon, I was required to have my arm plaster-casted from wrist to biceps with a 90° bend at the elbow. I have to say that I was somewhat stunned by his recommended course of treatment and the prognosis. Then, in double-quick time, combination punching, I was put on sick leave and my posting to the Saudi desert cancelled. That was as simple and as final as that. No one can gainsay the doctor. I had witnessed real power being exercised.

The speed with which the fates realigned my immediate future left me slightly disorientated. I was not confident that the treatment was proportional to the amount of pain I was experiencing, but the MO had referred to a medical text book, plucked from his bookcase, once I had outlined my symptoms, and doctors never lie, do they? I had no doubts that a few unsympathetic cynics would be preparing white feathers to present to me, but if I was honest, I would have admitted that, actually, I was pleased. No prospect of the boredom and discomfort sitting in a tented HQ, frying during daylight hours and alternately freezing during the hours of darkness. I would not feel the pressure of trying to perform in a role about which I was clueless and, most importantly to me, I would be on hand, unquestionably, to be a part of the son-of-BRIXMIS organisation slowly coalescing in the minds of the generals.

In the event, my analysis proved to be 100 per cent professionally flawed, and the war was fought, albeit on land, for a mere 100 hours. I could probably have cuffed that and bluffed my way through the test, even past so astute and cerebral an officer as the general. But I had missed out on a great personal and professional experience. I now regretted that I had not been a part of Desert Storm so I turned the other way, refocussed and carried on engaged in the business, closer to home, of hitting the Russians hard and maximising our intelligence in this critical period of European history.*

* Nearly a dozen years later I was serving as a military intelligence liaison officer (MILO for short), deploying as a troubleshooter with global remit out of the Old War Office in Whitehall. I had deployed to Albania – the same occasion that took me to the war graves in Tirana – before the British contingent arrived in theatre at the war's end. I had been there over three months, basing myself out of a flat in Tirana and the British Embassy, and driving north on specific tasks to the Kosovo border. When it was time to come home, I flew back to London in a 32 (The Royal) Squadron BAe 125. The only other passengers were General Rupert Smith and his US Army military aide, a full colonel. I had come across a small family carpenter who made traditional brightly coloured *Hasi* Albanian wooden cribs and had bought one from him intending to use it as a flower trough. It was a pretty and distinctly original souvenir of happy times. In order to protect it en route, and particularly whilst forcing it into the cabin through the over-wing port of the 125, I had wrapped it in my Army Arctic sleeping bag. As we settled down for the flight and strapped in, General Rupert pointed to the great green maggot resting against the rear bulkhead and asked me what I was doing with an Albanian cradle. How on earth did he know what was shrouded in this amorphous khaki mass? He knew because he is an extremely clever, intelligent man. It was one of those rare occasions when I was truly lost for words.

17

Rebirth

The last BRIXMIS tour returned back over the Glienicke Bridge on 2 October 1990. In fact, eleven tours returned over the Bridge that evening. With the Chief included, this was the full complement of on-pass tour officers tacitly paying recognition to the far-sighted work of Robertson and Malinin all those years ago. This final symbolic deployment of BRIXMIS personnel into the old DDR, teetering on the edge of independence with one day to go to the reunification of the two Germanies, was to celebrate the end in the traditional way the Mission said its farewells: with a barbeque. The G-Wagens passed through Potsdam in convoy to rendezvous at the pre-arranged site somewhere in the local area, where no doubt the wake was fully celebrated by all, tinged by a manfully restrained tear in the eye of a few, and sombre hearts in others. I suspect the occasion, in many ways, was akin to that of the funeral of a long-sick relative. Death had been anticipated for so long that, when it finally came, it was met with preparedness and even some relief.

That evening at the Mission House the Union flag was lowered for the very last time and presentations made to the officers of SERB, their job now done too. There was some trepidation and uncertainty on the part of some BRIXMIS members as to what the immediate future would bring, but nothing as daunting as the prospect facing our Russian friends, who watched powerless as fate stripped them of their superpower status and their state began imploding as it fractured into its constituent republics. For them, the immediate, middle and distant future looked truly bleak and not a little frightening as the country slid towards economic and political darkness.

The following day, the German reunification process culminated in the Day of Unity celebrations as one country's flag disappeared from its pole outside 405 East 42nd Street (the New York headquarters of the UN), whilst its bigger brother grew by 108,000 square kilometres and added 18 million people overnight to the tax register. It was an emotional occasion on both days.

The Bridge had been packed to greet the returning tour vehicles, not quite to the extent of the million-plus people who caroused in front of that now familiar party ground between the Reichstag and the Brandenburg Gate the following evening, but enough to demonstrate that something of historic proportions was unfolding. As the darkness thickened on 3 October, the sense of euphoria and once-in-a-lifetime occasion again possessed the monstrous crowd as it yelled and danced and clapped almost as vigorously as it had done a year and a month before as this epoch-making journey had begun. Again amongst the crowd, I tried to capture the sheer size and exuberance of the revellers, taking shot after shot with my small pocket camera, the same one I had used that evening thirteen months before. But as the counter reached forty, I knew there was an issue with either film or hardware. I could not believe it when I confirmed my worst suspicion. For the very first and last time in my life, I had neglected to load a film into the camera. What a time and event for such stupidity and carelessness – 'Surely it must signify something,' I tried to console myself!

It, 3 October, was very much a day and an event that justified that greatly overused cliché, 'the end of an era'. For the first day since September 1946, no Mission tour was on task in eastern Germany. Certainly this strand of the tripwire was now demonstrably severed. Of course, another was about to replace it and would remain in place, strong, taut and just as alert. The Missions themselves did not formally cease to exist until New Year's Eve, but slowly and surely staff began to drift away on posting, and for those that lingered, waiting for the end, the time resembled the rundown to the holidays as the long summer term petered out.

But, to quote another cliché, 'Nature abhors a vacuum.' And so, as the Missions were swept aside by the turbulence and fickleness of realpolitik, they were replaced, in the British case, by the Joint Intelligence Staff (Berlin).

The staff already existed as a small cell and had been responsible for updating the commander of the Berlin Garrison since it had been established at the end of the Second World War. But now, under the leadership of Hamish Norrie, my old boss from the FRU in Lisburn, it subsumed what cynics would label the rump of BRIXMIS, but what I would of course parry, describing us as the cutting edge of the last knife left behind on the rack. In reality, those of us who remained were a very strong contingent of ex-Mission operators.

The exact terms of reference and the operational directive for the new unit remained opaque to me. Frankly, I was not particularly interested in the small print so long as we got out early onto the ground and proved that the new modus vivendi was both practical and effective, and that there was still value in working on the other side of the now defunct Wall. I am, however, of the firm belief that we, JIS, existed on the personal initiative of the then commander-in-chief of BAOR, Sir Peter Inge, or 'Sir Inge' as the Americans referred to him – innocently or with humour, I'm not sure. This would certainly explain the wafting odour of inter-service rivalry that we occasionally detected seeping under our still-locked front door, and explained the dramatic exposé as 1991 unwound that one of the RAF officers in the new team had been pressured by a senior Royal Air Force commander, back in the Rheindahlen HQ, to report discreetly, and disloyally – without mincing words, to spy – on JIS activity.

Whatever the internecine politics, Hamish was the best officer to lead the team. The position obviously required professional courage, leadership and operational acumen, and he had all of these qualities in spades. Our overarching mission was to continue clandestine reporting on the Russian forces still garrisoned in eastern Germany, despite reunification and the Moscow government's firm commitment to redeploy its forces in the Western Group back home to the CIS, once Gorbachev had dissolved the USSR in March 1991. Obviously, the *Nationale Volksarmee* had now dropped off the target list – there was a NVA no longer.

Two tasks fell out of this mission: we were required to continue to collect intelligence on new weapons systems and equipment entering service in the east, no differently than BRIXMIS, and to record routine military movements and activities, with particular focus on the areas that had formerly been protected by PRAs – we did not give credence to any protective measures the Russians may have invoked, or continued to do so. More pertinently, we were to monitor the outflow of equipment being re-roled back home. This was particularly relevant to our German allies, who did not possess the capability to maintain any form of revised Russian census, and who were paying the Russian government a per capita sweetener for every truck, tank, big gun and plane removed from their soil. Surprise, surprise, there was a very strong suspicion that, of course, the Russians were grossly inflating the numbers and scale of the withdrawals, ripping off the German treasury.*

When I left in 1992, JIS was a team of fourteen operators, including Liz M., an Int Corps sergeant who became the first British female soldier to tour in the old DDR. Nine of these were former BRIXMIS operational staff. Four of us had served in Northern Ireland on Special Duties, including the CO, Hamish, who had served with both FRU and 14 Company. As a small team, we had a disproportionately heavy hardened steel nucleus – real hard-hitting experience and finely tuned professional expertise. On the demise of the Mission, the unit had been larger but as time went by we lost more and more former BRIXMIS originals as they returned to their own units, but we did have a number of new operators posted in so we were able to maintain more than a sufficiently sharp operational edge to get the job done. The big, and key, difference between us and the Mission was how we got that job done.

JIS deployed covertly in plain clothes and we used a mixed fleet of West German-registered civilian cars that we leased, actually from the NAAFI. Since the border with the Federal Republic had opened up, the east was awash with 'westies' in their cars, demonstrably lording it over their 'eastie' cousins, so there were few real operational security issues resulting from our western tagging. We did not stand out. Team-wise we dropped from three 'tourers' to two to make more efficient use of our smaller team, and to reduce our visual profile to third-party observers

* Early on in our new operational calendar we were visited in our offices, still located on the old Mission corridor, by a trio of *Bundeswehr* officers, attached to the German foreign intelligence service, the BND, as was certainly then common practice. They were part of a small team of around ten or a dozen, if I remember correctly, that were now working with the US team, designated as CAD-B (Combined Analysis Detachment-Berlin) that had succeeded USMLM. With the small talk over, we got down to hard business. They wanted to know how they could set up and operate a clandestine section mirroring JIS. Of course we helped them and gave them a comprehensive and tactically sharp brief. Either we did a poor job or they did a worse one, as we later heard that soon after our meeting, one of their team – possibly one of our visitors – had been shot and badly wounded climbing over the wall of a Russian barracks. I guess that may effectively have shut them down as an extremely convenient adjunct to US operations, potentially leaving us as one of the few *de visu* bean counters out on the ground, but I know CAD-B certainly continued in business.

– I've always followed the idiom that two's company and three's a crowd anyway. 'Two-up', we functioned as tight, cohesive teams and did not look like the *Stasi* 'Alfies', threesomes in their Ladas and Škodas of the 'old days', as we journeyed throughout the new West German lands. For me, if I was not operating alone, this was the ideal way to tackle any task. There was no one who I dreaded spending a long day with, and I certainly enjoyed being exposed to some new bands and genres of music as we all brought along CDs to help pass the time on the long journeys we made transiting from Berlin to targets and back. We now coincidentally mirrored more closely the rationale of the former US and French Missions. We reduced tour lengths from three days to single-day tasks, which meant that when we covered more distant targets, especially in the deep south and south-east, they were long, hard days, even more so in the harsh weather of winter. And we began to hit the Russians far more aggressively than BRIXMIS ever had. I found myself out on tour usually three days per week for the duration of my time in the unit, which without doubt represented a formidable operational pace.

I cannot confirm what our former colleagues in the French Mission were doing; I imagined pretty much what we were. We did not cooperate at the working tactical level or openly share intelligence in any systematic way – at least that I was aware of, and I think I would have been – with either them or CAD-B. These were days governed more by national interests and priorities, by bilateral relationships with the German government and the commitment to keep the latter up to speed on the Soviet withdrawal. Frankly, I preferred it that way.†

In addition to CAD-B, the US had set up at least one large reception centre in the west of the city to process 'refugees' from the east who wanted to be fast-tracked into western Germany. The intelligence personnel working here, including US, were on the lookout for 'travellers' who had played roles in the old regime and were of intelligence interest. When they were identified, they were debriefed – this was just another strand of HUMINT collection and certainly was a symptom of the turning wheel of history as, as a strategy, it had paid dividends in the immediate post-war years of de-Nazification. No doubt a small percentage ended up making different

† With the closure of USMLM, I had no further dealings with any of my old American friends or colleagues.

journeys to the one they crossed into the west to make. I visited the centre a couple of times, just to get a feel for what was going on there.*

We retained all the BRIXMIS tour-related files, folders and tour preparation materials. Following the guiding maxim about not fixing things that remain perfectly serviceable, we retained the proven functional organisation that had been a key characteristic of the Mission's success; thus, the corridor remained reassuringly recognisable. So, we employed the services of a slimmed down Special Section to process the 35mm films we took – why reinvent the wheel when one can inherit the nuts and bolts of a tried-and-tested and, frankly, winning team? We also employed the services of a weapons officer, Crispin, a lovely guy who I had known in the Province. Stinky Spandau remained to process our Russian documentary finds, as we would crucially still be TOMAHAWK-ing, only now during the hours of daylight as, working in plain clothes, we were no longer bound by the niceties of operational security that had ruled the Mission routine. The Research office remained in place, continuing to collate our intelligence, and an ops staff remained to task JIS touring. The ops room was run by the former BRIXMIS assistant ops officer, who had replaced Gary, commissioned from the ranks of the Royal Tank Regiment. He was an extremely experienced 14 Coy operator. In terms of other 'hard hitters', we retained the services of Keith from the Regiment, utilising his common sense and practical approach until he was recalled by Hereford. He was always easy and very pleasant company, and very effective and efficient out on the ground – he possessed that quiet calm of the West Country and the no-nonsense attitude that I believe marks out his former unit, the Royal Marines. There was no one in the small focussed unit who did not contribute something to the party and work to make JIS's contribution, amid the final death throes of the Cold War in Germany, a positive one. Only Geo failed to survive the transition and, frankly, that was because, quite simply, it no longer had a justifiable function.

We achieved a great deal in a relatively short space of time and with minimal resources. There is no debate that JIS had the best possible start, commencing ops on the sound springboard the Mission had constructed

---

* The West German government had always been processing asylum seekers from the east in a reception centre they ran in Marienfelde. This continued to be operated post-Wall until reunification. The centre then re-roled, interestingly, processing ethnic Germans seeking to return to the fatherland from Russia.

and fine-tuned over all those years – we remained mindful and thoroughly grateful for the operational heritage that, as the son of BRIXMIS, we inherited. In repayment, we wrote and closed the final chapter in its history, I think displaying professionalism, commitment and bringing home substantive results, even though this achievement is virtually unknown and unrecognised – but that is not important. We recognised and so did those who needed to know, I think, that we were an important player in the final act as the Warsaw Pact received the *coup de grâce*, and the West celebrated final and total Cold War victory, the prize being the lasting peace in Europe – the Balkans and Ukraine aside – that has endured to this day.

Rolling our sleeves up, we knew that equipment, once the withdrawal started in January 1990, was being expatriated to the motherland either by rail across Poland, a less and less attractive route with the new democratically elected government installed there, or out by sea. The most frequently used route by sea commenced in Rostock, but the ports at Mukran and Saasnitz were also used less regularly.

Consequently, the location I visited most frequently during my time with the new unit was the port in Rostock. Prior to joining JIS I had never even visited the city. Change, novelty and variety is always welcome in my life. It was a three-hour drive one way from the Olympic Stadium – of course, Potsdam, the Bridge and the old House no longer had any operational significance for JIS. Nor did we pay homage to the Americans by dropping in and dropping off tour highlights. I did miss the fridge though. We just raced north and west and south and east as required, by the quickest *Autobahn* routing.

I have always believed that if you display enough confidence and an air of indifference that suggests one belongs, you can get away with blue murder. This was how we got, repeatedly, into Rostock's port complex. In the early days of our touring here, we had been concerned that gaining access to the loading quays might be difficult. We did not want to have to work more remotely from OPs, as this would have been operationally problematical. In the event, it was not and, for us, there may as well not have been a main gate. Driving boldly through it, I was never once stopped to be quizzed as to my presence or to produce any identification. I expect that rapidly our regular and protracted loitering was noted and then regarded as normal, and I suspect that, living under the flag of the new Germany, the port staff knew the aim of our presence and, if not positively supporting it, at least grudgingly tolerated it. I suppose if access had been an issue, a route into the German government would have been identified and a formal arrangement established. We were, after all, indirectly working for Herr Kohl and the people of his new *Deutschland* – and ostensibly for free.

Once in the port complex, we quickly learned which quays were used for on loading onto the transport ships and ferries and positioned ourselves so we could discreetly observe the hardware as it was prepped for shipment – on busy days, it was patently obvious. Somehow, the presence of wheeling, cawing seagulls was incongruous, set against our touring background in the Mission, which had been predominantly inland, and gave one the feeling that we were in another country that, in reality, pedantically, we were. In this way, we sat for hours and recorded thousands and thousands of individual moves as the normal routine of port activity carried on around us, at least pretending to be oblivious.

Reversing my former cynical stance, I now appreciated that we had to log every piece of hardware as it was moved out of the country. Every VRN and side number contributed a word to the pages of the unfolding saga. We were very much in the business of recording the departure of every battalion, of every regiment so that we could strike it from our master orbat and label it as 'Returned Home'. Beyond that, it was actually extremely interesting to observe the routine of a busy working port, the massive cranes unloading other 'normal' cargo and all the other various to-ings and fro-ings. It brings me back to my comments in *The Deadly Game* – I have a sneaking regard for logisticians. It is they who make our transnational world go round.

We did not score any major scoops, as no new major equipment was observed being out-shipped, but equally we did not experience any security incidents or challenges as we conducted our protracted observations. I found it always stimulating to be breathing in crisp, salty sea air. I found energy levels would rise surreptitiously and moods would brighten. I would also experience a sense of imminent discovery and bubbling excitement. I do not know whether this positivism was triggered simply by memories of childhood holidays at the seaside or, as I prefer to think, by something primevally imprinted on my DNA as an islander, as we Brits are. Whatever, it served as a powerful combination of emotion and drive to accompany an operational task.

I chalked up a couple of good sightings in the hinterland, away from the coast, including a line of SA-11 launchers on rail flatbeds, obviously heading eastwards. Had we still been with BRIXMIS, I might have got a pat on the back – in JIS, the pace was too fast and furious to worry about trivia. Our attacks on the static locations that we inherited from BRIXMIS fundamentally proved to be, as far as my memory stretches, relatively routine and unremarkable in intelligence terms. But what we were able to report on systematically was the continued pull out of Russian forces as barracks were vacated and left to the vandals and scavengers. This was positive negative intelligence if one was to engage in Rumsfeldian-like word acrobatics. Not exciting, but a crucial role underpinning our very existence.

One thing that became readily apparent was that we missed the robustness of the Mercedes G-Wagens on marginal and sub-standard surfaces. Modern saloons are wonders of technology, and ours were marvellous at effortlessly digesting the distances we routinely covered on our uni-day tasks, but they are not designed for rough, tough, uneven forest tracks or the sandy cross-country of heathland, let alone the sticky mud of a winter's day. And we still had to periodically tackle some challenging terrain to access the less accessible locations we needed to check. I was bogged in a couple of times in situations that the G-Wagen would have scoffed at: once in sand, which was relatively simple to extricate ourselves from with some improvised ersatz planking and a quick lowering of tyre pressure, and once, more seriously, in a sucking quagmire that I had underestimated as we gingerly traversed it. On this occasion, self-help was hopeless, and I had to smile as I flagged down a passing UAZ-469 and its crew, who very kindly pulled us out with absolutely no clue as to who they were aiding.

After some very rough and uneven work way down south in the area around Chemnitz, formerly Karl-Marx-Stadt, we damaged the underside of the car and had to call in at a small local garage for some impromptu running repairs to guarantee we would make it back home. That turned out to be a very long day. Once we had returned to base, I had a heated conversation with the ops offr, trying to impose some sort of surveillance-type finger-pointing debrief that I had witnessed on my surveillance course. He criticised me for damaging a vehicle. I was not taking any such nonsense and we ended up almost toe to toe – use a kitchen knife to cut down an oak, you will get the job done, but you will have problems with the continuing functionality of your knife. The cars we deployed were not designed for harsh off-road treatment – if we had the tasking, we had to be prepared to take the concomitant hits. Our relationship never returned to the disinterested neutral status quo upon which it had carried on previously. It was only the second time in my whole twenty-seven-year career that I remember abandoning calm, reasoned logic and the pursuit of conciliation, losing my temper. I'm sure he lost as little sleep over the discord as I did and, having set a new tone, we continued to clash on a number of other operationally related matters, all kindergarten stuff in reality, before I left the unit, in my mind, job virtually done.

One of my favourite locations to tour was a small town far to the south in the beautiful state of Thuringia: Rudolstadt. The town sat on the edge of the Thuringian Forest, conjuring images and sensations of its Teutoburg cousin to the north-west where an alliance of Germanic tribes in 9 AD, taking advantage of the heavy forestation, had destroyed Augustus' Roman army and turned the tide on the city state's advances into Europe.

Elements of 79 Guards Tank Division were barracked on the edge of the town. To the west, a crescent of high ground provided the perfect vantage point to sit in cover and look down into the garrison, watching the comings and goings as its inhabitants went about their business, not quite ant-like, oblivious of the stalkers in the hills above them.*

I visited the site, with both the Mission and JIS, and remember the weather always being bright, sunny and warm. These are some of my most enduring memories of working in the DDR, somehow comforting reminiscences, peaceful with their sense of relative detachment. It was almost akin to the contentment sitting at a pavement café watching the world going by, a part but not a part of everyday drama unfolding – the familiarity and security of the soap opera – as we sat looking down on the ways of man like some Greek god on Olympus. Intelligence gleaned from these visits was minimal, but again the presence of the normal justified the effort. Years later I was unavoidably transported back, and revisited these remembrances, when we had an au pair who had grown up in the town – just one of life's small, and acceptable, coincidences. Sadly, she turned out to be useless.

Fortunately, despite the efforts of others, I have never had to deal with any serious injury or wound sustained by a unit member, or to myself. I did, however, attend to the victim of a head-on car crash in the middle of the former East Germany during one of our long days out.

---

* Interestingly, Steve Gibson also highlights Rudolstadt, in his book, as one of his favourite target locations. A case of great minds, or fools never differing?

A small western car, I think a VW Golf or Jetta, had crossed the barrierless motorway central reservation and ploughed headlong into the front of a GAZ 66 lorry coming from the opposite direction. The car was registered in Hamburg in West Germany, with its distinctive 'HH' number plate prefix. The driver must, presumably, have been one of the flood of visiting businessmen.

Cars and trucks continued to flow uncertainly past the wreckage. This was still the East – no one wanted to get involved if they could avoid it. I did not need to share their understandable and conditioned reticence. Potentially, I could help. I pulled onto the hard shoulder. Controlling the natural human instinct to rush to the scene, I waited for a break in the traffic before carefully, and speedily, crossing the carriageways. The scene was a mess. The truck driver was unscathed but looked to be in a state of shock, standing beside the cab of the truck. His Russian-made GAZ was largely undamaged – the front was supposed to be flat. Not so the capitalist-made machine. Clearly he, the truck driver, was not at fault and appeared to bear no responsibility for the crumpled wreckage before him. The bonnet of the VW was twisted metal; were it not for the VW badge on the tailgate, the make would have remained guesswork. The car's engine was off, I guess as a result of the impact. Broken glass was everywhere.

There was a single occupant inside. He looked about 30, short mid-brown hair, his round face a pallid and bloodless thing. I could see teeth on the dash and clinging to the front of his unbuttoned jacket. He was wedged in place, still seat-belted but at an angle across the front passenger seat. There was a cassette tape protruding from the player mounted in the dashboard. Somehow, I knew instinctively that he had been focussed on changing the tape, probably removing it to change sides, had lost control and veered out of his lane and across the carriageway into the path of the unsuspecting trucker.

The door was jammed shut, so I leaned in through the shattered front passenger window and, inserting my fingers into his mouth, cleared the driver's airway of tongue and bloody mess – these were definitely risky days as the AIDS epidemic gathered pace, but I did not feel any fear, let alone squeamishness, that prevented one from coming into contact with this poor young man's blood or any other bodily fluids. I checked for any external bleeding as best I could and ascertained that he still had a pulse. I felt pretty powerless but fundamentally unfeelingly detached as I watched his laboured breathing – this was not any lack of humanity on my part but a coping mechanism I find that I am able to switch on and off when confronted by extreme situations. Undoubtedly, his severe head injuries were compounded by internal damage, which probably included his chest and resultant injury to his lungs caused by the seatbelt.

I wondered who he was, what his story was, how his day had started and what his concerns, worries or expectations for it had been as it unfolded. I wondered whom he was missing and who was missing him back home, the other side of the old Iron Curtain, in Hamburg. I wondered if anyone had had any premonition as to how the day would end, his final day, cut short in a pointless accident on a nondescript stretch of motorway in the middle of nowhere, in a land that had ceased to exist. Whatever injuries he had suffered proved so severe that he died within minutes, still wedged inside his car before the East German ambulance arrived. Nothing could have saved him that day.

Psychologically, we faced an interesting and ambiguous situation as we sallied out into the new east. *Our* mental state was positive, well founded and balanced, but I detected a degree of uncertainty and ambivalence on the part of the remaining Russian forces. I am not sure the conscripts fully understood what was happening politically around them, although I am sure communication was better than in the days of the invasions of Hungary and Czechoslovakia, when their forebears had had absolutely no idea of where they were and what they were engaged in. But what I thought I detected was a lack of confidence and surety when we watched them and began to probe their security a little more aggressively than we had in Mission days. They were no longer the kings of their eastern castle, and they seemed to sense, if not explicitly realise, it.

On one occasion, a cold inhospitable day, I was out on foot, in plain clothes of course, and was close in to the outer external fence of a barracks, just searching for a vantage point to see the tanks we expected to see, or indeed their absence, suggesting they had already started their journey home. Whoever I was out with was in the car, securing it and awaiting my return. As I continued my walk along the track that bordered the fence line, a Soviet sentry, behind me, challenged me in Russian, ordering me to stop. Perhaps he shouted that he would fire if I did not, but this level of subtlety was beyond my grasp of the Russian language. I did not react to, or acknowledge, his presence or challenge in any way. I did not hear any distinctive cocking of a weapon – perhaps it was made ready already – and I carried on strolling away, casually, ignoring him and his AK-74, shoulders hunched, head bowed, minding my own business. Nothing happened – no follow-up, no attempt to stop me, no further challenge. The sentry was confused and unsure of what to do next. I was confident this would not have happened in Mission days had I been there in my green uniform, readily identifiable as an enemy soldier. That degree of enhanced scrutiny, best case, would have invited immediate detention and the whole merry dance with the *Kommandant* et al. or, worst case, a Nicholson-like shooting.

Out on the dry, sandy tracks of the heathland of the Letzlinger Heide, we had been idly following a tank transporter that we had run into as we searched for activity on the training area, enjoying the access that had been denied to BRIXMIS though all those long years of PRAs. Atop the low-loader was a T-80. It was tricky to determine whether it was en route to or from an exercise, or maybe just requiring a trip to the command workshop for repair, so we kept behind it, maintaining a discreet distance. We were free running. This was an uncanny instance when my comment made from the cockpit of the Chipmunk regarding barracks security and the contrast between their 'fronts' and 'backs' bounced into sharp focus.

We maintained our station behind the transporter and swung gently to our left further into the dipping glare of the low-lying sun as the state of the rough tac route started to improve. I reached down into the door for my sunglasses but they were still in my bug-out bag, where I had put them after our last stop. It was in the car's boot. I squinted. Surreally, before we knew it, there were one or two buildings, and there were soldiers and some women and kids milling about. It was an odd and slightly startling transformation. I glanced quizzically at Simon. He was looking equally quizzically at his map book. We were amongst a few more scattered assorted figures. We were not only deep in what had been a PRA but, chillingly, the realisation dawned on me that we had strayed and were now equally deeply slap bang in the middle of a small barracks. We had come in through the open backside, straight off the training area – no warning, no security, no formality of any degree.

'Bollocks, bad tradecraft and bad situation awareness': self-recrimination that I decided to keep to myself.

But I did kick myself metaphorically, hard. I mentally blamed tiredness for this lapse in concentration.

The tank transporter had pulled up and stopped. I had to make a snap decision and quickly as we passed it, incidentally confirming the side number of the T-80: 'Should we stay or should we go?' As the words of the song enquire. Shockingly, in front of us, just 300m away, was a gate, the main gate to the installation, and we were the wrong side of it. There was now considerable activity around us but it was benign, certainly at the moment. We must have been drawing some attention but, equally, our presence must have appeared fundamentally non-threatening and 'normal'. We were there, so we must have belonged there, right?

My decision was made. Act normally; act with confidence.

'Hang on, Simon.'

I continued easing the car forward, low speed, low gear, high concentration. Forward we went, towards the gate. Time began to slow; my vision and my perspective began to narrow. The gate was closer … closer, closer, closer. We were now upon it – we were looking at a bloody big metal double one with a big Soviet star cut out in the middle of both the left and the right halves. It was rusty, it was old but it must still work. Things were now going to go either very well or very badly.

The moment of truth winked at us as the sentry began advancing towards the car from where he had been lounging, leaning against the gate he was guarding. He was approaching my side and looking expectantly towards me, the driver. I stopped the car, feeling naked and exposed, knowing full well that we were sporting West German number plates but conscious that this was knowledge that we alone likely shared. This was no time for flakiness. This situation required grasping and the projection of positive confidence. I engaged the sentry eye to eye, animal to animal, and imperiously pointed with my extended first finger – no time for finger confusion – at the metal barrier. He looked at me, looked at the gate, then back at me. He appeared momentarily uncertain. I pointed again and nodded reassuringly at him. I could see the wheels turning as he computed the situation. He performed the required half turn and moved purposefully towards the large L-shaped bolt. I was top dog. I had won this minor battle of wills – thanks be to God.

With his AK-74 slung over his right shoulder, he drew the bolt and put his weight against the right-hand half of the heavy barrier. I eased up the clutch and began to roll forward. No time to stall the car. I gave him a stern-faced nod, raised my right hand in salute – and tacit thanks – and eased over the threshold, turning left onto the main road beyond. Absolutely nothing to it. I resumed normal breathing and smiled at Simon sitting next to me, a signals officer unluckily posted into the Mission on the back of his Russian language training, just as the doors closed for the final time.

'Where to now, Simon?' And we headed off to our next target. Sometimes there is just no need for more than the barest minimum of words.

I always liked coming out with him as his wife always made him better sandwiches than the ones I made for myself, and he would always swap one or two.

We had been lucky. But we had kept our cool and reinforced our right to be there – even if, clearly, we had had absolutely no excuse to be the wrong side of the fence – had acted confidently and pulled it off. The sentry, and the rest of the inhabitants, had been fundamentally unsure of who we were, and no doubt uncertain if action was required, but they had concluded that, as we were there, we must have been authorised visitors. In these changing, fluid days, they had gone with the flow – the path of least resistance has a draw all of its own to humankind no matter background, culture or heritage.

Another day, another dollar, I had been out on tour with the SNCO, coincidentally another member of the Royal Signals, who had taken over as the 'manager' of the Mission House just before it had been handed back to our Russian former landlords. JIS had kept him on too for man management and family reasons rather than seeing him reposted away like an unwanted parcel.

We were out of the car on foot, making a quick survey of the edge of this small training area, looking for any signs of rubbish tipping, either documentary or in terms of discarded equipment, as we assessed that the unit was soon to move back to Russia and would no doubt be looking to slim down its profile. All had been quiet up to now and uneventful. As we crested the slight rise in the terrain, we were met by an instantly sobering sight, not that we had been drinking anything other than uninviting flasked coffee. A half company of Soviet soldiers in column, two abreast, rifles shoulder-slung, was marching steadfastly towards us up the muddy, uneven track. There were approximately fifty of them, perhaps a touch more, and they were roughly 70m away and closing. They had, of course, seen us as we had seen them – there was nothing else on this tree-less, desolate heathland to draw the eye as we crested the skyline. The miserable day had just worsened.

Now, if I had been on the high street of my local town and saw someone approaching from the opposite direction whom I despised and loathed, or who bored me to death, or both, I might have considered a quick about-turn to avoid having to acknowledge them. This, however, was not Ashford or Aberfeldy or Aberdare or Antrim high street, and a dramatic reversal of direction, no matter how tempting, would only draw suspicion, if only from the officers leading their conscripted company, and invite possible confrontation and incident.

Out of the corner of my mouth, I told Don we would continue our direction of travel, unchecked. We were now about 40m and still closing. If there was a point of no return, we had certainly already passed it. We were committed. Whatever potential protection wearing a Mission uniform might have afforded, this was not such a situation. This was always the case when conducting intelligence-gathering operations in plain clothes. Abandoning the uniform of the soldier labelled one fairly and squarely as a 'spy'. Spies received scant sympathy. JIS did not enjoy the recognition and, therefore, was not subject to the implicit code of conduct that governed these relations that BRIXMIS had benefited from. But this was why JIS could do things the Missions could not have.

We could now start making out individual faces. They were the usual rough-looking bunch, a collage of the disparate facial features that made the old Soviet Union such a fascinating experiment in diversity and inclusion. We were eyeball to eyeball now, clearly able to see the whites of their eyes. We maintained our stride and nodded at the front rank as they drew level. As we passed down the line, I noted irreverently that somehow our position was akin to the general inspecting a parade of men, as we made fleeting eye contact as they snaked past us. It was an amazing sight and a very special experience. Here we were, the closest I had ever been to our former enemy, outside of one-on-one conversations at social functions or the impromptu brush contacts with reggies or other small isolated groups of Sovs out on the ground. We were so close we could smell their unwashed bodies and soiled uniforms, see every stain, scuff and tear on their well-worn khaki winter jackets, every missing button, every badly shaved chin and the lank unwashed hair of their hasty cookhouse haircuts. They were indeed a rough crew. They looked seasoned and as tough as any inner-city gang from the roughest, toughest estate, not the press-ganged youngsters that, in reality, they were. There would be few other scenarios or occasions where others would share our experience that day as both the Soviet Union and the Red Army marched in step towards oblivion – that was quite a sobering thought.

We reached the end of the column. They marched on; we slouched on. I risked a backward glance. Not a jot of any lingering interest on their part. A perverse experience. We were their enemies and they were our enemies; we had been for forty-five years. On the command of our leaders, we would have done everything in our power to kill each other, yet we were perversely able, literally, to rub shoulders and amble on, all of us, peacefully towards lunchtime. For me, although I was oblivious to the reality then, this was the very last time I would be 'hand to hand' with the Soviet bogeyman.

Again, we had displayed the confidence of belonging. They had been unsure of who we were and their own jurisdiction. It was bad luck that we were wrong place, wrong time, and we could not have anticipated or avoided the encounter. But we had got away with it. I silently hoped that life back home would be tolerable for the rest of their time in uniform and, on demobilisation, the economic and political uncertainties of the new Russia would not be too hard on them.

OP TOMAHAWK continued with the enhanced quality of product that operating under daytime skies guaranteed. We were running what felt like a completely different operation. Being able to see clearly and move more confidently and safely reinforced how tough TOMAHAWK-ing had been in Mission days. Being able, with a new clarity, to spot and assess the potential of documents and gear in broad daylight saved so much time and effort and ensured a quantum improvement in the ratio of real product to mere stinking junk. The excitement of a 2 a.m. foray had now gone and the true unpleasantness of what we were doing was clearer to comprehend, but we were only ultimately interested in results. These were better days.

The Mission had always found that the odd East German would turn up on a dump at the most perverse hour of the night but rarely caused much problem; we found now that we were in much fiercer competition with an increased number of locals now things had opened up, but this too was generally not an issue. Obviously, we were after distinctly different prizes. However, we did have to be cautious not to highlight our seemingly perverse interest in non-commercial resaleable rubbish and disguise that what excited us were items that could not be recycled, but were military-related detritus.

On one occasion, having come across two large anti-tank portable launcher tubes lying there amongst other rotting garbage, we found ourselves unable to retrieve them because of the intensity of activity of our fellow scavengers. We did not want to expose our interest in what were clearly two weapons and of no interest to any sane civilian. They were awkward and heavy, but we moved them as casually as we could and discreetly hid them, covering them with other garbage, returning very early the next morning to get them into the back of our Ford estate and return them to Berlin for analysis whilst the rest of humanity was rightly still tucked up in bed.

On a similar theme, I was out with Tam, a Russian-speaking Int Corps SIGINTer who worked in the Spandau office. He was a solid, bright Scotsman, always good company and a professional to the core. He had been with me when we had found ourselves bogged down in our quicksand. This time, it was midsummer, 1991, and we were making our way along a track running alongside a line of trees. As we passed one of the generously spaced trunks, I noticed something hanging from one of the lower branches. I stopped the car. Tam had spotted it too. Without a word, he jumped out of the passenger side, approached the tree and, reaching up, unsnagged this unorthodox fruit.

I could see clearly now that it was a small shoulder-fired anti-tank weapon, suspended by its khaki carrying strap until Tam had freed it. What a weird, bizarre place to find a discarded weapon. It made little sense. In another scenario, I might have thought about a booby trap, but not here on the outskirts of the village of Nowhere-dorf, literally in the middle of nowhere. Tam checked the weapon was in a safe state, a reflex SOP, and brought the tube back to the car. Neither of us were subject matter experts but it seemed unremarkable and looked like any other short portable anti-tank weapon – its significance, however, was that neither of us could attach a name or nomenclature to it. I opened the boot and Tam placed it carefully inside and covered it with the hessian we routinely carried.

Back at the office at day's end, we handed it over to Crispin, the weapons officer. We were rather startled at the enthusiasm of his reaction. Apparently, we had brought home a new weapon for the tech int community to examine and reverse engineer. It was, incredibly, a hitherto unidentified and new mark of portable shoulder-held anti-tank missile launchers. We had our scoop, if that was what either of us had been striving for. Ironic that it should have come so late in the history of the Cold War. Word later percolated back from the techies that this was a significant find and we were to be given a pat on the back – sometimes, apparently, intelligence does grow on trees.

At the height of the Cold War such a find would have had significant implications. It might have indicated a potential technological advantage in the Soviets' offensive anti-armour capability and revealed a gap in our own defensive posture. Our revelation bore no such import as Russia would no longer be facing off against NATO on the European plains, but as an export model, the weapon could be used against Allied forces on any potential developing country battlefield, conventional or asymmetric.

Whatever the big-picture significance, the find was built into the unit brief that Hamish delivered to high-price visitors. It was all grist to the mill.

Paradoxically, but not unexpectedly, despite our freedom and ease of movement, sadly our human source operations diminished in scope and scale. This was a great pity because, with the opening up of the old DDR, it had become so much easier to physically conduct operations. However, the fundamental problem, as I have already explained, was that with no official status akin to that which BRIXMIS had enjoyed, and our clandestine status as JIS, we were almost completely neutered when it came to attempting to engineer points of contact whereby we could get alongside potential or actual Soviet targets and begin the process of establishing relationships.

We did, however, manage to meet a couple of officers, encouraging them to cross into West Berlin to meet with us, complete with my favourite common operator counter-surveillance phase in the orders. But with the goalposts now repositioned so far from their original Cold War locations, there was little utility conducting meets at this operational level – as targets, they just did not have access to the calibre of intelligence that the community was now demanding and, rudely, expecting. Frankly, from our perspective at the operational sharp end, we ourselves did not need tactical-level intelligence relating to an army that was in full 'retreat' either. How the battlefield had changed. Collective sights had now been refocussed on the state of affairs within Russia, where the palpable sense of turmoil and unpredictability, combined with enduring menace and capability, demanded attention. To use language we did not use then, for us it was essentially game over.

The relentless routine, however, continued without incident or respite. We watched more and more equipment, and more and more units of men, move homewards as Russia maintained its commitment to end its forty-seven-year occupation. We confirmed the closure and abandonment of more and more barracks and gained entry to have a final rummage and check for anything of value carelessly left behind. And I received the first speeding ticket in my life, ironically being stopped after passing through a sneaky *VoPo* trap armed with a new capitalist speed gun, a few kilometres an hour over the limit. It exemplified that the regime was already in a state of advanced decomposition and that it had, with unholy haste, willingly or as the consequence of institutional coercion, assumed the values and priorities of the Western world – I had to laugh. I showed the *VoPos* my military ID card, my MOD 90, which, arguably, put me beyond their jurisdiction as we still did not recognise any organ of the eastern state, but took their ticket as a souvenir. Having lived in Switzerland, I don't laugh as speed cameras are so commonplace and pernicious there – talk about indirect taxation. Watch the Rowan Atkinson *Johnny English* film in which he is on a mission and driving through a small Swiss village. It lights up like a town being carpet bombed as he passes through the twenty speed cameras positioned on its winding main road. Welcome to the Helvetic Republic, Johnny.

Having already been picked up for my major, I now had a posting order for June of the following year, 1992. I was excited to be going to the Manor as head of Surveillance Branch, responsible for surveillance training, double-hatting as the 2IC of SIW. In the event, I assumed responsibility for Research Branch as well, agent-running training, taking on the title of training officer. It was not an operational job, although an operational deployment to Bosnia would emerge from left side during my four and a bit years there, but I would still be very much involved in the Special Duties world. For training to be effective and relevant, there had to be a strong and established link to the operational 'customer', so I would be forming a close relationship to all the Int Corps' operational units, not just in the Province but in Hong Kong, Belize and the other theatres within which we operated. It would also be nice to be home, in England, and have the opportunity to see a bit more of my family and make new friends.

Inevitably, once you know what the future holds in terms of one's next role, your sights unavoidably rise imperceptibly and time can start to drag a little. I would have been in this amazing city and been a player in the far-reaching events that had burst open the international scene, seemingly from nowhere, for three years. It was probably the right time to go and, anyway, we were close to that rare point in military life when there was nearly no job left to do.

But there was still a full life to live and an endless, almost breathless, span of social experiences to greedily consume. One work-related incident that I will not forget, however, concerned the Moscow coup in August 1991 when rogue elements in the government and KGB came close to ousting Gorbachev, sealing Yeltsin's future as the hero who would assume the reins of power once he had engineered Gorby's retirement in the December. Naturally, in Berlin we closely monitored the events and probably had a broader, better grasp of the situation than most.

Coincidentally, I was on a short, sharp visit to the HQ in Rheindahlen on JIS core business. I was cornered on several occasions by members of the staff there and quizzed about events as they unfolded in Russia. On one of these occasions, my interlocutor was the visiting deputy director of the Int Corps, a senior officer who I had always liked and knew well from FRU days in the Province. At times, he appeared somewhat absent minded and dithering – this was not the case but it could give people, who did not know him, the wrong impression. Whilst we stood there, as he quizzed me about the coup, it was one of those days when he appeared stressed. In response to a question, I gave him, in my opinion, a full and well-considered answer. As he appeared to struggle to process my outpouring, his retort was, 'Don't confuse me with facts' in his slightly stuttering way. It was one of those moments – we both ended up laughing at this very un-intelligence officer-like response. It underlined his humanity and lack of pomposity despite his position and seniority. It has always stuck in my mind and continued to endear me to this good officer and man.

The intensity of operational life in Berlin was fully matched by its social life, both in Mission and JIS days. Life was an endless round of eating out, rock concerts, theatre, cinema, ballet and opera, if you chose to grab these unique opportunities and, frankly, had the time. I did grab them when I could. Tickets, even in the west, were cheap, public transport was so good and the city was relatively small and compact that all venues were within easy reach. All the top rock tours took in the city, whether they performed in the wonderful outdoor Waldbühne, the Olympic Stadium or the host of smaller scenes. I saw everyone from Midnight Oil to Prince, Billy Joel and

The Beach Boys to Sinéad O'Connor and The Waterboys, to Jason & the Scorchers and Gianna Nannini. To my eternal regret, I missed The Rolling Stones in the Olympic Stadium. With the Wall down, I spent the day at a brilliant open air concert in the east of the city enjoying Tina Turner, Die Toten Hosen and a host of supporting acts.

Once the Wall was down, following the New Year's Day run, as Mission members we were permitted to cross into the east of the city, where previously whatever agreement had paradoxically forbidden us entry, to enjoy the high-quality opera, ballet and classical concerts in prestigious venues like the *Staatsoper* located at the top of the Unter den Linden. Initially, we had to attend in uniform, in our mess kit, with champagne and canapés during the interval. These evenings were very real and special occasions. The performers were all Eastern Bloc but world class. Later, things relaxed and we would cross more freely and unobtrusively in civilian clothes.

One of the choices I received the most pleasure and satisfaction in return from was my decision to play my last two full seasons of rugby. I had played on and off, more off than on, during previous tours, having the most fun with the FRU team that Hamish had got together in the Province. Cheekily, we had ordered a stack of white Fiji rugby union shirts for games. Opponents were slightly confused about the significance of the palm tree motif and any implication of an association with the botanic gardens in Belfast but I'm sure failed completely to spot the relevance of the acronym.

I had looked at the garrison side representing the units up at the Olympic Stadium but decided I would take a more unusual and original option. I decided I would try out for the Berliner Sporting Club (BSC) and play in the German league. It was a path I did not regret taking for a single moment. A peripheral factor I have to admit that swayed my decision was the location of the club's training ground and venue for home fixtures. It was literally 200m up the road from where I lived, near the top of the city's famous Kurfürstendamm and directly adjacent to my favourite local restaurant, which doubled as the unofficial club house. I also liked the story that during Nazi times, BSC had been the only sporting club in an otherwise devout city not to have sworn allegiance to the party.

The standard of rugby was perhaps not great, and certainly below that of the garrison side, but the BSC boys were deadly keen and fit. From my perspective, it was flattering to be a big fish in a small pool and I did not have to fight to keep my place out on the lonely wing. We played all the other sides in Berlin, including all the military teams in the city, and then once the Wall had been breached, we travelled out not only into the east of Berlin but down to Halle and Leipzig and other 'unorthodox' venues – I had not realised that, by then, the German rugby union was already over 100 years old, so we were a part of a much greater tradition.

I had got hold of a DDR military calendar featuring poster boys and girls from the NVA, the DDR's Air Force and Navy. Mr March was a steely eyed, chisel-jawed paratrooper straight out of central casting – the very epitome of the fierce, viceless, incorruptible defender of the socialist ideal. He was semi-concealed in brush, his AK-74 shouldered at the ready to despatch a capitalist NATO aggressor. The next time I saw him, he was in the pack of the *Lokomotive* side from a Berlin workers' suburb. We had a post-game laugh when I pointed out this association to him.

On my 30th birthday my speed left me literally overnight. It was as stark as the accelerator cable in a car being cut. But I played on and enjoyed the pantomime of being in the team that played a student side in Potsdam. At this juncture in the reunification calendar, I still had to travel over the Bridge in uniform, get changed, play, lose, then re-change into my lightweights and woolly pully, complete with BRIXMIS badge, for the after-game drinks. I later heard from Sergei Ostroumov that SERB had considered playing a practical joke on me, as they were aware in advance of my request to come over and play rugby. They had discussed turning up at the end of the game and staging a mock detention, no doubt taking me off to the SERB HQ for a drink. I think, on reflection, I'm glad they abandoned the idea but maybe it could have been the basis for a unique article in *Rugby World* magazine. I somehow appreciated, and was not a little flattered by, the fact these hardened Soviet intelligencers had sat down and considered something fun and light-hearted in contrast to the seriousness that their role and the times represented – it portrayed the human face that I had observed many times constrained by the Cold War mask that we were all slowly learning, gradually, to lower.

As I entered my final year with JIS, I organised a ten-day adventure training exercise in Sardinia, taking advantage of RAF transport between the NATO airfields of RAF Gatow and Decimomannu. I remember being accused by some staff officer in the Berlin HQ of cutting corners, circumventing red tape and bending established processes and procedures to finalise the plans and make it happen. But somehow I felt the team deserved a break in the heavy schedule that we had accepted in monitoring the most significant troop movements in recent world history, underscoring the relaxation in Cold War tension that had guaranteed the continued existence of my 'friend's' rules and regulations. I took his comments as praise. I think we, just as our antecedents amongst all those past generations of Mission tourers, believed in making things work and getting the job done no matter the restrictions, the hazards and the challenges we all had confronted on the way. This was an enduring aspect of undertaking Special Duties and living and working in full view of the 'normal' Army – you were subjected to, and judged by, a set of standards, rules and expectations that were largely not relevant or appropriate to our roles and to the missions we tackled. But it got the team away for a break, providing some non-Russian focus as the bulk of equipment of the mighty Russian bear that had crouched poised, outstaring the NATO alliance for those four decades, had already shipped out and the pace of withdrawal had already slackened to a trickle.

Then it was suddenly all over, like a band finishing their last song and promptly leaving the stage. It was time to leave, mission accomplished – absolutely no reference to George Bush's similar words – and with brave new worlds both to leave and to re-enter. Incredibly, peace had endured. The Cold War was won and lost without the wholesale slaughter and destruction that had pulled like a choking dead weight at the collective psyche of both east and west for so long as a future that appeared so predictable, so inevitable and so terrifying. Job well and truly done.

18

A Last Glance Back Over My Shoulder

Whilst this book is not about the allied Missions' cousins, the Soviet Military Missions operating in the western half of the old divided Germany, they cannot escape a brief mention because the wider structure, that BRIXMIS was a part of, critically included them. Because we were enemies, inevitably we cannot get away from measuring their impact and success against our own. Based in their secure compounds in Bünde, Baden-Baden and Frankfurt, what they speculatively appear to have achieved must be factored into any assessment of the overall relative successes of the whole Mission concept.

Their continued existence was, as I've already suggested, an integral feature of the quid pro quo that allowed the AMLMs to operate for as long as they did. Certainly, the US high command settled into a ready ambivalence, effectively happy to turn a blind eye to their Frankfurt guests – keeping their head in the sand regarding its activities until the very last breath – because their view of the Cold War stalemate from their HQ in Heidelberg was that, fundamentally, they needed the access into the Red Army that USMLM gave them above any other consideration. The Americans remained oblivious of the real collection and operational missions of SMLM-F, and the bigger-picture penetration they suffered at the hands of KGB and GRU agent handlers and the agents they ran – if London and Paris were perhaps less sanguine, this is a conclusion about which I am insufficiently informed to support or deny, although I have been reliably informed that UK Plc had a more informed handle on at least some of the activities of SOXMIS.

Identifying the highly damaging Soviet penetration of the USAREUR staff suggests the reality that the whole Military Mission playing field was not an even one, and that the six individual Missions were not playing the same game. Whatever balance the Allies felt – or even hoped – existed was almost certainly not real. The sparse evidence that has been exposed to open scrutiny suggests that, at the very least, the Soviet Mission based in Frankfurt, in the heart of the US sector, played a more potentially lethal role, and one far wider than the US Command assessed.* It is reasonable to extrapolate that these insights equally identify the likely operational activity of the two other Red Missions. Whatever, they provide a teasing appetiser to what may yet still be tossed into the public domain when the appropriate number of years have passed and sealed classified archives are opened – if ever.

Fundamental to, and underpinning, any look at the SMLMs, is that assessment that – compared to Soviet access to the West, particularly West Germany – in terms of the relative ease of intelligence collection, the Western allies needed the unique and unequalled access into the former DDR that the post-Second World War establishment of the Military Missions guaranteed, so long as the Malinin series of bilateral treaties remained unchallenged, unquestioned and in place. There was deemed to be no other way that quality intelligence collection against Warsaw Pact forces could be mounted with the same reach and comprehensive clarity as the Missions provided. Their product was unparalleled *de visu*, up close and personal intelligence. This remained the case for most of the longevity of the Cold War until national technical collection means negated some of this utility. Upon the back of this reality, the strategic decision was taken, and adhered to over the years, that nothing should jeopardise this access – nothing should be allowed to challenge or potentially curtail the Western Missions' freedom of movement.

This led to the decision to minimise countermeasures against the AMLMs' counterparts in the west that might provoke reciprocal sanctions

* There is nothing I have been able to identify that makes reference to the activities of the Soviet Mission operating from Baden-Baden, nor of French operations against its personnel. The British effort was more joined up and long running but remains classified; however, at least some US counter-intelligence perspectives are detailed in open source, most comprehensively by ex-US Army counter-intelligence officer Aden Magee.

directed at the lodger units in Potsdam. The consequence, especially on behalf of the US, as Magee describes, was scant attention being paid to the small Soviet unit based in Frankfurt. This ostrich-like approach was not even mitigated by the exposé in the Rezun/Suvorov book – easily missed though it was – identifying that the three Soviet Military Missions were regarded by the Russians as full-blown GRU residencies operating in parallel to their diplomatic stations in Bonn and Cologne.† Half-hearted attempts were made to keep the Soviet Mission compound in Frankfurt under static surveillance to monitor touring patterns and then, much later, to attach remote surveillance devices to Soviet tour vehicles – resulting only in the likely compromise of US technical surveillance capabilities when they were undoubtedly discovered – but these were not sustained or robust counter-intelligence operations.

A significant factor in the US counter-intelligence approach to the potential threat(s) posed by SMLM-F was that after 1954 – before which touring was more aggressive than that undertaken by the AMLMs in the east – the Frankfurt Mission began to maintain a touring profile with just enough punch to mirror its Allied counterparts, who were aggressively targeting the Soviet orbat of GSFG/WGF, and technical targets, full time. It is assessed that this was a tactic designed to disguise what had become the GRU tour officers' primary operational roles – and put quite crudely, Moscow knew everything, especially about US forces in Europe and their war plans and, in reality, did not view mission touring as an intelligence source. This change in operational tempo resulted from the GRU and KGB securing strategic-level intelligence from other sources, certainly including at least those highly placed human agents, as already mentioned in Chapter 6.

Compared to the AMLMs working out of Potsdam, Soviet tours were short, less than a working day, and after taking account of transit times to and from Frankfurt, amounted to just a few hours 'on target'. This created a catch-22 paradox: because tours were deemed to pose a low threat to US forces based in West Germany, they were paid scant attention by counter-intelligence assets. This lack of collection further reinforced the initial assessment that SMLM-F was not conducting aggressive hostile intelligence collection missions. The self-licking ice lolly was further

† *Inside Soviet Military Intelligence*, first published in January 1984.

substantiated by a comparison of Mission size. Numbers of touring passes were identical, as determined by the clauses of the Huebner-Malinin Treaty, but compared to the seventy-strong staff of USMLM in West Berlin, the Frankfurt Mission numbered only fourteen. This small team, so the assessment went, was simply not capable of posing a significant or dangerous threat, so ran the flawed, tautological logic. By reaching this decision, early on, not to take measures against the Soviet equivalent that might risk limiting USMLM's freedom of movement, the West, especially the US Command, had trapped itself.

A revised, more informed judgement was not made until 1986, after four decades of hands-off lethargy, after the creation of the Det A surveillance team, born out of the DIA's Foreign Counter-intelligence Activity (FCA), based in Fort Meade, Maryland.

If the assumption is correct that at least the Frankfurt-based Mission's primary focus was not on mounting orbat or technically related intelligence collection tours against US Army Europe units, or as a liaison tool as laid down in the post-war treaties establishing them, then what was its primary mission?

In the absence of hard open-source intelligence in the public domain, one can only speculate and turn to the few published readily available accounts of Soviet intelligence operations that provide potential explanations.* There had always been some speculation in some quarters that the three Soviet Missions were involved in supporting agent operations. But beyond this, what the defectors, detailed below, revealed was a whole different ball game. They asserted that Soviet intelligence was caching weapons and other equipment in western Europe, particularly in West Germany. It is likely that, as GRU units, the SMLMs played a key role in these clandestine operations. Deeply concerningly, these caches included, so the defectors stated, small low-yield tactical nuclear devices: 'suitcase bombs'. In *The Cold War Wilderness of Mirrors*, Magee speculates that perhaps the Soviet Missions had a responsibility to monitor their state, if indeed they had not actually conducted the initial 'caching', and to react when their reserve battery systems kicked in. Were this to have been true, then the SMLMs

* Three such books – in addition to those by Suvorov/Rezun – by former KGB officers that raise intriguing scenarios are Oleg Gordievsky's *Next Stop Execution*, Vasili Mitrokhin's *The Mitrokhin Archive* and Stanislav Lunev's *Through the Eyes of the Enemy*. Lunev also testified before the US Congressional Select Committee on Intelligence.

were not solely intelligence units but had crucially important operational roles in preparing the 'battle space' for World War Three in western Europe, and may even have been given tasks related to deploying these weapons if and when tasked. Perhaps in due course, this speculation may be confirmed through official sources, but I sense this may be an outburst, from Moscow, of honesty and openness too far. The intriguing question is, where is all this cached equipment now?

The decision was finally taken to address the imbalance and to task and deploy Det A, the DIA's premier surveillance team, to watch SMLM-F. The team spent the initial months of 1989 engaged in intensive work-up training in preparation for commencing systematic operations, including urban ops, against the Mission.†

Following the year-long negotiations that followed the death of Nick Nicholson, the Soviets had conceded an additional 35 per cent of unrestricted territory, formerly PRA, in the DDR that the Allied Missions were now able to tour freely within. What they sought in return was acknowledged and unfettered freedom of movement in the major conurbations, particularly within the American sector. This included downtown Frankfurt. It is arguable whether the true significance of this piece of horse trading was appreciated at the time, but whatever the degree of consciousness, the decision was taken to focus surveillance on what had in the past been regarded as seemingly routine and unimportant Soviet Mission activity away from the barracks and training areas against which USMLM and the British and French Missions so actively and aggressively targeted in the east.

By April 1989, Det A was deemed ready for operations. The inaugural 'downtown op' was paradoxically set for 15 November. Of course, in the meantime, with the breaching of the Wall, the world changed, history changed step and the operation was postponed.

Det A was, however, deployed against routine tours beyond the confines of Frankfurt and during the first half of 1990 identified a garage in Kassel

† In the early 1990s, a team of us from SIW in Ashford went over to Fort Meade to train and exercise with the FCA surveillance capability. They were very professional and clearly had benefited from previous Manor visits that ensured we shared an identical methodology, right down to the curt radio procedure used by the operators out on the ground.

engaged in fuel coupon fraud.* Members of SMLM-F were identified passing a large quantity of coupons that they had been issued, but had not used, in a deal whereby the garage kept half of the value of the coupons on redemption and the Mission members kept the other half – a relatively substantial sum.

It was not until 20 August 1990, on the back of a three-week exercise, that the first follow was authorised to get behind SMLM-F members deploying inside the city. The first three days of the operation proved uneventful, but on 23 November, the deputy head of the Frankfurt Mission – who had already served a number of tours in the Frankfurt Mission, plus an attaché tour in Washington DC – departed the Mission compound in a SMLM-F-plated car, with driver, and headed towards the city centre.

Once the vehicle had joined the dense traffic of the Bornheim district, it abruptly halted and the colonel swiftly alighted as the driver pulled away. The surveillance team spotted the quick stop and dropped foot operators, transitioning smoothly from mobile to foot surveillance, the sure sign of a good team. They tailed him to the Paulskirche, a famous landmark church in the city. En route, wearing nondescript barrack-style uniform without hat or rank insignia within the spirit of the Malinin treaties, ensuring the Soviet officer blended unobtrusively into the crowd, he was seen to execute rudimentary and cursory anti-surveillance drilling. This was no doubt the consequence of the belief and continued assessment, based on years of precedent that, as a member of SMLM-F, he was not under hostile US surveillance.

He was followed into the church by a fA operator and spotted sitting next to an unidentified male. It was obvious the pair were talking, but because the surveillance operator could not communicate openly with the rest of the team, for fear of being compromised, when the second man moved rapidly back out of the main entrance of the church, he was lost to the team deployed outside and never subsequently identified. This was the first substantive evidence that US counter-intelligence had of what clearly appeared to be a brief agent meeting involving SMLM-F officers.

Within 30 minutes the deputy head of the Mission had been picked up by his driver and was back inside the Soviet compound.

---

* Allied forces stationed in West Germany received fuel coupons, effectively halving prices at the pumps. As the Soviet Missions were administered by the Allies, they too were entitled to an issue in a reciprocal arrangement whereby the Allied Missions could draw cheap fuel in the DDR. The system was tightly controlled but was still open to fraud.

Later that day, the order to conduct urban operations against SMLM-F was rescinded. The East German parliament had issued a declaration earlier that day that the DDR would unite with the West German state.

Like their Western counterparts, the Soviet Missions operating within West Germany were disestablished just over a month later on 1 October 1990. On the face of it, they packed up and left, leaving even less of a mark on German political, cultural and social life than we did in the east, but I have little doubt their impact on the Cold War intelligence landscape was far greater and more sinister. The brief insight into their activities revealed here, at the end of their four-decade endeavour, suggests very clear focus – at least in terms of agent handling operations – but, with little to allow us to gauge the tempo, intensity and targets of their operational activities, just how much damage, real and potential, was caused to western interests can only be guessed at. To be pessimistic, it is worth reiterating the fact there were three Soviet *rezidenturas*, all steered, one must assume, under unified GRU control in terms of operational tasking – their objectives would have been ruthlessly and aggressively focussed, and complementary across the three Missions, the length and breadth of West Germany.

In contrast, over the Wall, the Allied Missions – despite their gritty professionalism, their tenacity and creativity, their sheer hard and painstaking work and application, their acceptance of unrelenting long days and short nights when touring, the harsh summers and even harsher winters as unrelenting backdrops to each and every tour, the uncomplaining night-time hours spent rummaging blindly through putrid, pulsating rubbish dumps and the unthinking courage pushing hard against every type of target, constantly under the watchful eye of the *Stasi* and too many trigger-happy Soviet and East German sentries, perpetually looking for one more scoop or sighting to answer that one last open and unresolved question fed into the Operations and Weapons staffs in the Missions – were the national assets of three different countries, let alone competing agencies. And one of them, the French Mission, was not even working under the North Atlantic security umbrella. Ultimately, despite the cooperative framework that shaped touring tasking, each worked to its own agenda, percolating down from their centres of government.

It has to be my personal conclusion, formed now thirty years after finally leaving and locking my office door in the Olympic Stadium for the last time, that although we, without debate or doubt, won the war – the

Cold War – I am, however, fearful that, despite the stellar achievements of BRIXMIS, USMLM and FMLM, in terms of battles, we may very well have lost the 'Battle of the Missions'. The Soviet KGB and GRU penetration of NATO military staffs and the compromise of the Allied Mission collection efforts, and the clear, though limited, open-source indication of strategic-level tasking of the three Soviet Missions operating in West Germany – quite possibly aiding the penetration of those same espionage targets – suggests we were out-trumped and outmanoeuvred. Western thinking, especially on the US side, was so intent on protecting their own collection capability beyond the Iron Curtain that they essentially ignored the Missions they hosted. Whether the British and French attitude was more balanced will perhaps one day become clearer – on the British side, with at least the release of intelligence derived from the view from inside the surveillance cars of 28, their eyes on SOXMIS.

Perhaps, if a fuller picture of the roles and activities of the Soviet Missions emerges before too much more water flows silently under the bridge, it may be possible to more definitively re-evaluate the outcome of this battle, perhaps painting a less pessimistic conclusion. Only then might we be able to more objectively define the part they played in the totality of the wider war that we, in the West, won, finally drawing a balance between us and them. I'm not holding my breath.

Regardless of this, I am still in awe of the contribution made to Western security, and indeed the security of mankind, over all those years of fear and nuclear dread, by that tiny band of men and women, the Commander-in-Chief's Mission, that will always go by the name of BRIXMIS. I am proud to have been privileged to play a very small bit part in the history of probably the most remarkable and successful unit in the history of intelligence and espionage.

As Brits we love to talk about 'punching above our weight'. BRIXMIS did just that, round after round. It was *our* enemy who ultimately threw in the towel.

19

The Omega: The End

It had been an incredible experience, across the whole spectrum of activity and emotion. As I look back at this rich treasure of memories, I know that I was so, so fortunate and privileged to have been able to experience them. We had seen a new volume of European history – in fact, world history – open and been right there on Page One as minor fringe actors in Chapter One. And we could not forget that, just as significantly, we had been on stage as the curtain had lowered on the last scene of the final act of Part One – to have been over the Wall, 'behind enemy lines', across the bridge, unknowingly as the build-up to the fall of the Wall gathered its silent momentum, then, with Gary, to have crossed back again under the wire just hours before this momentous event shattered all our realities, was unbelievable, but unimpeachably fact.

The world, however, has certainly not become a safer place, nor a more predictable one, but we had witnessed the beginning of the end of the terror of the Cold War and its mutually assured ability to destroy the planet. We had seen the death of one sovereign state, but the rebirth of a now united people and the creation of a new superstate at the heart of Europe – for better or worse. We had witnessed a staggering outpouring of emotion, and the first demands for the dividend that always accompanies the perception of a new peace. But for us as a military, we just changed step and marched on, leaving one set of challenges to face emergent new ones. And I, and my former colleagues, would not be idle.

In its turn, JIS checked that the family had definitely moved out, that the Russian House was finally empty, ensured all the windows were shut, switched off the last lights and closed and firmly locked the front door. I don't remember what we did with the key but hopefully it will not be needed again despite the sabre-rattling and cynical brinkmanship of Putin's Russia – armed adventurism in Georgia, Crimea and now Ukraine proper, notwithstanding.

I made my return trip back home to the UK following the same, but now very different, route I had driven only three years previously, but what, in real terms, could have been thirty.

Peace in Europe, the Balkans and those western fringes of the old Soviet empire apart, did not translate into peace across the globe as the lifting of superpower pressure allowed oxygen to seep in and fuel the tensions and stresses that had hitherto largely been repressed and contained by the balance of Eastern and Western Bloc forces. However, these conflicts, although they brought untold suffering to combatants and innocents alike, did not, and do not, yet, threaten the very continued existence of mankind. The risks we face today and tomorrow, including those posed by a meddling Iranian theocracy, hopefully soon to be brought to book, not least to Israel and the people of Palestine, will no longer exclusively be founded, predominantly, on interstate competition and aggression. Instead, they will be predicated on fault lines that are mainly, but again not exclusively, the catalysts of demographic, environmental and cultural pressures – the most dangerous cultural conflicts will erupt where the plates of heterogenous civilisations clash. Where state actors engage with each other, the battle space they will increasingly deploy in will be a cyber one, although, particularly in East Asia, it is highly likely that clashes will embody very real, but probably restricted, kinetic elements.

They, as risks to master, whatever their genesis and nature, will not require armies to respond with classical massed military means, nor will they, I suspect, demand the endeavour, commitment and resolve of an equivalent USMLM, an FMLM or, for that matter, a BRIXMIS. History does not always repeat itself …

Annex A: Debrief of a Stasi Officer

Introduction

1. Subject is a thirty-year-old former lieutenant in the MfS. Despite having joined the Stasi in 1978 he was not fully employed until just prior to its demise, after a protracted period of probation and training. At the time of the dissolution of the MfS, subject was employed in the small eavesdropping unit based in Potsdam, targeting Allied Mission phone calls into and out of their Potsdam Mission Houses.
2. He readily agreed to be debriefed about his knowledge and experiences within the Stasi. His motivation is largely ideological based in that he now sees clearly the faults and excesses of the old regime and the part played in it by the MfS. He now feels no loyalty to his former organisation or its masters and wishes to 'purge his soul' in an almost confessional way.

Recruitment and Early Life in the MfS

3. Subject's ambition from the age of 14 was to join the NVA as an officer specialising in air defence. Students pledging themselves to the

NVA enjoyed secondary education up to degree level. Subject made contact with the schools liaison organisation in Potsdam.

4. At age 17, however, he was approached by a Stasi officer who worked out of the Wehrbezirkskommando in Potsdam. He agreed to join the Stasi in preference to the NVA. Subject met the officer several other times whilst still at school. In this way he learned more about the Stasi while still not understanding in which department he would work. The Army was not happy on learning that one of their candidates had been poached by the MfS.
5. On 01 Oct 1978 he formally joined the Stasi. He attended a 2-month basic training course in Glienicke, north of Berlin. This consisted entirely of basic military skills. Students received no special-to-arm intelligence training. On the last day of the course a car picked up subject and took him to Potsdam, to Beyerstrasse. This is where subject was employed, in the eavesdropping unit, on the dissolution of the MfS.
6. In 1981 he was asked if he was prepared to learn a foreign language. From April 1981 to November 1983 he studied English at the School of Foreign Languages in Schönwalde, based in buildings next to the castle. All students were members of the MfS (some were candidates for the Foreign Intelligence Service) as were the teachers. The majority of teachers had served abroad.
7. In November 1983 subject returned to Potsdam, to Beyerstrasse where he was introduced to the eavesdropping section.
8. In 1986 subject began a course at the Hochschule in Eiche in Potsdam for management positions. During this period students were attached to other departments to gain on-the-job experience. For example, subject spent 6 months attached to a surveillance team. The course covered political indoctrination, international law, penal law, and more operationally related skills including clandestine photography, recruiting and running agents. The subject completed this course early in 1990 just prior to the Stasi's disestablishment.
9. Subject's work in the eavesdropping section was a prelude to his ultimate transfer to the analysis cell, also based in Beyerstrasse. Its function was to sift through the raw data pertaining to Allied Mission activities and identify factors that could be exploited.

Perception of Role of Allied Missions (AMLMs)

10. The MfS perception of the AMLMs was that they were intelligence-gathering units closely linked to their national intelligence organisations. Not all Mission members were believed, however, to be professional intelligence officers.
11. The major tasks carried out by the Missions were perceived to be:
    a. Collection of military data, primarily Soviet-related but including NVA targets. This included ground installations and deployments, radar and communications sites, airfields and aircraft. When targeting aircraft it was believed the AMLM vehicles could intercept air-to-ground communications. It was also believed that seismic sensors were used to detect the movement of ground equipment.
    b. Collection against certain economic targets of strategic interest eg the uranium mining operation in the Weimar area.
    c. Recruitment and running of agents in the field. The running of agents was believed to include direct contacts eg receipt of tactical tip-offs from Deutsche Reichsbahn employees concerning rail movements, or reports from factory workers, and indirect contact including the servicing of DLBs.

Mission Operations

12. The effort expended by the MfS against the AMLMs involved several different departments or 'Referat' using different means. Monitoring the Missions involved mobile surveillance teams on the ground, telephone eavesdropping of Mission House calls and the use of Unofficial Members' (agents) sightings eg shopkeepers, petrol station attendants, residents near military installations. The whole anti-Mission operation was directed from Beyerstrasse in Potsdam. The fundamental problem was that so much raw information was collected that the effective processing of it was not possible.
13. The MLM unit worked to Abteilung 5 of HauptabteilungVII in HQ MfS in East Berlin. It consisted of 7 Referat employing approximately 85 people:

   a. Referat 1. Ran all Unofficial Members working against the AMLM including the Mission House workers.
   b. Referat 2. Worked to develop awareness of AMLM activity amongst the East German population. They also liaised with HQ MfS on matters of army counter-intelligence.
   c. Referat 3 & 5. Deployed the surveillance teams.
   d. Referat 4. Included the analysis cell and eavesdropping section.
   e. Referat 6. Oversaw all admin matters.
   f. Referat 7. Controlled the surveillance teams on the ground using radio.

14. The aims of the MfS anti-Mission operation were:
   a. Prevention of espionage against military targets. It was perceived that the Missions mainly, but not exclusively, targeted Soviet military activity and installations. By harrying Mission vehicles with surveillance teams it was the Stasi's intention that they would be forced to abort efforts to gather intelligence on military targets. Sometimes the Soviets would provide the MfS with prior warning of new equipment entering the DDR in order to protect it. MfS would then mount an operation deploying surveillance personnel to the deployment area to prevent Mission attack. Such operations could last several weeks. Surveillance operators would live in tents or caravans in the area of the target and deploy from there. The team would be briefed on the basis of an appreciation by the Analysis Cell in Beyerstrasse. They would consider who from the Missions would be likely to attack the target based on perceived specialisations and how the target would be approached by the Mission crew based on routes, OP locations etc.
   b. Prove Mission espionage activities. Another role of the mobile surveillance teams was to photograph or video Mission teams photographing targets. The aim of this was to provide the Soviets with proof upon which to base official protests through SERB.
   c. Deliberate planned detentions. One of the main aims of mounting deliberate planned detention operations was to prove a direct link between the Missions and national intelligence services in the form of captured equipment or documents. The decision to

mount a deliberate planned detention – as opposed to opportunity detentions executed by troops on the ground – was based on one of two factors. If a Soviet Mission crew was detained in West Germany, the order would be passed to the MfS in Beyerstrasse by the KGB department in Potsdam to attempt a reciprocal detention. Alternatively, on the basis of analysis, if a member of the Missions was deemed to set patterns whilst touring that could be exploited, he would be targeted. An officer of the FMLM was successfully ambushed and detained because he persistently approached an identified OP from the same direction. A special squad was set up based in Zossen Wünsdof composed of either GRU or Spetznaz personnel. Because the Soviets did not have the analytical section to plan detention operations they relied on the MfS. Additionally, the special detention squad would be supported on the ground by MfS personnel dressed in Soviet uniform. The Soviets did attempt to continue planned detentions after the demise of the MfS in February 1990. Latterly, even prior to February 1990, little was done to detain Mission crews because of the complexity of mounting the attempts, the scant gain in reward and the reluctance to risk embarrassing the Soviet leadership in its dealings with the West.

d. Gathering information on Mission personnel. One aim of observation of AMLM personnel on the ground was to determine if they had any specialist skills or interests in any particular type of target. Uniform or insignia worn by members were disregarded. Personality files were produced by the Analysis Cell in Beyerstrasse on all Mission members and passed (unofficially) to the Foreign Intelligence Service by the Colonel and to ZAIG.
e. ZAIG 5 (Central Analysis and Information Group) was based in East Berlin and controlled by the Deputy Minister for State Security. Its role was to liaise with friendly intelligence services and in the case of Allied Missions to aid identification of ex-members who reappeared serving abroad in their national embassies or other agencies. Two ex-AMLM members traced in this way were Colonel Huet (an ex-French Mission Chief) who was posted to the French Embassy in Moscow and Colonel

Le Joy (ex-US Mission) who was posted to the On Site Inspection Group for Medium Range Missiles.

Surveillance

15. Surveillance operations tasked against the AMLMs were directed by the Central Department in Potsdam. There were two Referat surveillance teams, 3 and 5, one deployed in the north of East Germany, the other in the south. They were controlled by radio direct from Potsdam. When working against a Mission target the surveillance controller in Potsdam would relay information via the KGB to SERB and thence to the Kommandant in the area.
16. The aim of surveillance was to follow Mission vehicles clandestinely, monitor and attempt to record their activities on film or video. Their surveillance reports would then be used by the Analysis Cell to enhance future tasking.
17. A surveillance vehicle would normally have a 3-man crew, a driver, a navigator and radio operator. They had the capability to perform number plate changes in the field.
18. Formerly, teams would attempt to maintain surveillance on a Mission vehicle throughout the length of its tour in East Germany. This was, however, perceived as being too difficult. Consequently, tactics were changed and through analysis, choke points and likely targets were identified which surveillance teams could stake out and hope to get behind an Allied Mission tour.
19. To augment the efforts of covert mobile surveillance, the MfS operated an extensive system of static surveillance posts manned by Unofficial Members run by Referat 1. A number of people in the Seestrasse area of Potsdam were recruited to observe the French and British Mission Houses particularly when receptions were being held. No long-term OPs were maintained for fear of compromise. The Volkspolizei posts opposite each Mission House and the Soviet checkpoint on the Glienicke Bridge were tied into the surveillance effort. Each had a phone/radio link to Beyerstrasse. In this way a comprehensive all-informed net existed to monitor AMLM movements.

Telephone Tapping

20. The telephone eavesdropping operation mounted against the AMLM houses in Potsdam was conducted by 2 MfS departments. The technical installation and maintenance of the intercept capability was the responsibility of Department 26 based in Hegel Allee, Potsdam. The monitoring and recording of intercepted information was done in Beyerstrasse, Potsdam, part of Hauptabteilung VIII in HQ MfS East Berlin. Although each region had an eavesdropping section, Potsdam was tasked exclusively against the AMLMs.
21. The installation of the 'bugging' equipment was done by agents employed in Deutsche Post. They used an 'inductive' method whereby the actual telephone lines were not touched. This meant the Mission phone-users heard no telltale sign as they spoke. In all, 6 lines were 'bugged': 3 DDR Deutsche Post lines going out of the Mission Houses and 3 West Berlin lines coming into the houses. All 6 lines passed through Hegel Allee. They were then diverted and extended to the monitoring room in Beyerstrasse. Subject was not aware of any technical eavesdropping devices planted actually inside Mission Houses.
22. The operation in Beyerstrasse was technically unsophisticated. Latterly, cassette recorders were used to replace reel-to-reel machines, one per line. Each was voice activated when a telephone line became busy. The line in use was highlighted on an electric light board.
23. During working hours, calls would be transcribed directly from the telephone line by one of the 2 officers employed there, subject plus one. Once a day, important calls would be typed up into a daily report. During silent hours the calls were taped and transcribed the following morning. If important traffic was passed during the night the duty officer would call out a transcriber.
24. The daily report listing traffic was passed directly to the supervisor, a lieutenant colonel. He made 3 copies. One went to the KGB office in Potsdam, one to the Analysis Cell in Beyerstrasse, and one to the Colonel. He annotated to what end information should be used eg a separate report, an AMLM members P file etc.
25. All numbers dialled from the Mission Houses to West Berlin could be worked out. If they were not known they would be rung by

Unofficial Members in order to identify them and attempt to gather more information. In this way a number of Mission members' home phone numbers, mainly from USMLM, were known. Unless stated by the caller, the originating numbers of callers from West Berlin could not be worked out. The subject was not aware of any monitoring of telephones in West Berlin but he presumed another department was active in this field.

26. The tapes of recorded phone calls were kept for 2 weeks then wiped clean. Monthly and yearly reports were written by the Analysis Cell and kept for 10 years. With the dissolution of the MfS all the old reports were passed across to the KGB in Potsdam.
27. This operation was deemed, especially by the Soviets, to give some valuable intelligence, particularly of a HUMINT nature. Most was given away by USMLM. FMLM consistently maintained the tightest telephone security.

Penetration

28. The subject was not aware of any penetration of the Allied military including the AMLMs. In the case of the Missions, the Soviets would have run any such case. However, East German civilians employed in the Mission Houses as domestic staff were being run by the MfS's Referat 1 in Beyerstrasse, and the Potsdam KGB detachment as agents.
29. In the case of the British Mission House staff, they were certainly regular providers of the Berlin Bulletin. Copies were also obtained by the Colonel through his contacts in the Foreign Intelligence Service.
30. The only case of a recruitment attempt mounted by the Soviets against an AMLM member concerned a USMLM Warrant Officer. The MfS in Beyerstrasse was able to build up a large and comprehensive file on this man based on the extensive 'business' dealings he conducted from the American Mission House in Potsdam over the telephone. The Soviets were confident that they could exploit the Warrant Officer's business connections. The Americans became aware of the Soviets' intentions and decided to let the play run in

order to gain an insight into their methodology. The affair ended with 2 Soviet officers being apprehended.

Foreign Intelligence Service

31. Subject had no personal contact with the Foreign Intelligence Service. He did know, however, that the Colonel maintained unofficial links with them.
32. Subject is not aware of any of the identities of any Foreign Service personnel and was not approached by them at any time. He knew one officer who worked in a counter-intelligence department who was approached after ten years' service with a view to recruitment. He was eventually rejected because although he was married, he had a girlfriend.

KGB in Potsdam

33. The KGB detachment in Potsdam worked out of offices within the Soviet barracks on Friedrich Ebert Strasse, thought to be Potsdam 283. Access to the KGB building was carefully restricted. Subject never entered it. The detachment consisted of a military counter-intelligence section and a section of 10 to 15 officers monitoring the Allied Missions.
34. Although all contact between Beyerstrasse and the KGB was conducted by the Colonel, subject knew the first names of two officers, Yuri and Nicolai. They never used their surnames.
35. The KGB also had an officer working in SERB. Subject does not know his identity.

Demise of the Stasi

36. There was some optimism prior to December 1989 that the Stasi would survive the tide of change sweeping through the DDR. This was short lived. An attempt was made by reformers to occupy the

Beyerstrasse station. This was successfully deflected by the Colonel who persuaded the demonstrators that the offices were used solely for military counter-espionage work.

37. By the end of December 1989 all surveillance and eavesdropping work against the AMLMs had ceased. In January 1990 it became obvious that the pressure on the Stasi was too great and that the organisation would collapse. Whilst waiting to be told that their jobs were to go, the Stasi operators in Beyerstrasse caught up with paperwork, shredded material and generally found jobs to do to keep busy. The end was announced by the Colonel in February 1990. At lunch time the offices were locked up and everyone went home. By then, the station's entire paper holdings had been moved by lorry to the KGB office in Potsdam.
38. Since February 1990 the Soviets have attempted to approach a number of ex-Stasi officers in Potsdam, including the Colonel, but excluding the subject, with a view to persuading them to continue working for them, or at least to divulge information they require. There is a fear amongst this group that the Soviets, or indeed the BND, may resort to blackmail in order to persuade them. The Colonel point-blank refused to co-operate with the KGB and has advised those others who have been approached to do the same. He has also spoken to a number of Mission House domestic staff advising them what to do if they are blackmailed by the Soviets into aiding them.
39. In November 1990 subject again met the Colonel as he does periodically. The Colonel reported that he had been again talking to members of the KGB department in Potsdam. They claimed to be relatively inactive with little to do. It was expressed that hopefully a military coup in Moscow will not be long in removing Gorbachev from power.

Annex B: The Robertson-Malinin Agreement

AGREEMENT REGARDING THE EXCHANGE OF MILITARY LIAISON MISSIONS BETWEEN THE SOVIET AND BRITISH COMMANDERS-IN-CHIEF OF ZONES OF OCCUPATION IN GERMANY

In accordance with Article 2 of the Agreement of 'The Control Machinery in Germany' of 14th November 1944, the Soviet and British Commanders-in-Chief of the Zones of Occupation in Germany have decided to exchange Military Liaison Missions to be accredited to their respective Staffs in the Zone and to confirm the following points regarding these Missions:

1. The Missions will consist of 11 officers assisted by not more than 20 technicians, clerks and other personnel including personnel required for W/T.
2. The Mission will be placed under the authority of one member of the Mission who will be nominated and termed 'Chief of the Soviet/British Military Mission'. All other Liaison Officers, Missions or Russian/British personnel operating in the Zone will accept the authority and carry out the instruction of the Chief of Mission.

3. The Chief of Mission will be accredited to the Commander-in-Chief of the Forces of Occupation. In the case of the British Zone this means Air Marshal Sir Sholto Douglas.

   In the case of the Russian Zone this means Marshal of the Soviet Union Sokolovsky.
4. In the case of the British Zone the Soviet Mission is invited to take up residence at or near Zone Headquarters (Bad Salzuflen area).
5. In the case of the Soviet Zone the British Mission is invited to take up residence at or near Karlshorst or Potsdam.
6. In the case of the British Zone the Chief of the Soviet Mission will communicate with the Deputy Chief of Staff (Execution) Major-General Bishop or his staff.
7. In the case of the Soviet Zone the Chief of the British Mission will communicate with the Deputy Chief of Staff Major-General Lavrentiov.
8. Each Mission will have similar travellers' facilities. Passes of an identical nature in Russian and English will be prepared. Generally speaking there will be freedom of travel and circulation for the members of Missions in each Zone with the exception of restricted areas in which respect Commander-in-Chiefs will notify the Mission and act on a reciprocal basis.
9. Each Mission will have their own wireless station for communication with its Commander-in-Chief. In each case facilities will be provided for Couriers and Despatch Riders to pass freely from the Mission HQ to the HQ of their own Commander-in-Chief. These same couriers will enjoy the same immunity as diplomatic Couriers.

   Each Mission will be provided with telephone facilities in the local exchange for such communications (post, telephone, telegraph) as exists when they are touring in the Zone. In the event of a breakdown of the wireless stations the Zone Commander will give every assistance in meeting the emergency by providing temporary facilities on his own signal system.
10. Each Mission will be administered by the Zone in which it resides in respect of accommodation, rations, petrol and stationery against repayment in Reichsmarks.

    The building will be given full immunity.

11. The object of the Mission is to maintain Liaison between the Staff of the two Commanders-in-Chief and their Military Governments in the Zones. The Missions can also in each Zone concern themselves and make representations regarding their Nationals and interests in the Zones in which they are operating. They can afford assistance to authorised visitors of their own country visiting the Zone to which the Mission is accredited.
12. The agreement is written in Russian and English. Both texts are authentic.
13. The Agreement comes into force the moment letters have been exchanged by the Deputies to the British and Soviet Commanders-in-Chief of the Zone of Occupation in Germany.

(Signed)
B.H. ROBERTSON Lt.-General
Deputy Military Governor, CCG (BE)

M.S. MALININ Col-General
Deputy Commander-in-Chief,
Chief of Staff of the Soviet Group
of Forces of Occupation in Germany

Berlin, 16th September 1946

Annex C: Glossary

1RS: First Battalion of the Royal Scots (The Royal Regiment) amalgamated in 2006 into The Royal Regiment of Scotland but still recruiting in the Edinburgh area. Also known as the First of Foot, the First and Worst, and Pontius Pilate's Bodyguards.
2IC: Second in command.
14 Coy/Gp NI/JCUNI: All names used to refer to the Army's surveillance unit in the Province. Unit directed by Director Special Forces.
Agent: An agent, in intelligence terminology, is a human asset that is formally recruited and under full control, run by an agent handler from an intelligence organisation. As such, agents are taskable, steerable and are rewarded for the intelligence that they provide.
AMLM: Allied Military Liaison Mission. Included the Missions of all four former allies in both East and West Germany.
APC: Armoured personnel carrier. Can be wheeled or tracked. British examples include Warrior (tracked) and Saxon (wheeled); Soviet APCs include the BMP and BTR series.
ARBiH: The Bosnian Muslim army.
ASA: American Standards Association. Regarding photography, an arbitrary rating of film speed in terms of the sensitivity of film to light.
ASU: Active service unit. A terrorist unit.
ATO: Ammunition technical officer. A bomb disposal officer.
AWACS: Airborne warning and control system for coordinating air defence.

AWOL: Absent without leave. A military offence but less serious than desertion.
BAOR: British Army of the Rhine. The British Army units based in West Germany.
BfV: *Bundesamt für den Verfassungsschutz*. The federal German security service, in other words the domestic intelligence agency, equating to the UK's Security Service (MI5). It is based in Köln/Cologne.
BND: *Bundesnachrichtendienst*. German federal foreign intelligence service.
BRIXMIS: British Commanders'-in-Chief Mission to the Soviet Forces in Germany. The name was later changed to reflect the Soviet move from GSFG to WGF. Thus, BRIXMIS became the British Commanders'-in-Chief Mission to the Western Group of Forces. The Mission continued to go by the diminutive BRIXMIS.
BSSO: British Services Security Organisation. A Security Service (MI5) organisation responsible for liaising with German national and regional security services.
CAD-B: Combined Analysis Detachment – Berlin. The US organisation that succeeded USMLM. It forged a working relationship with the BND, initially using its operators to deploy on the ground.
CASCON: Casual contact. A human source but not under full control. In strict terminology, CASCONs are not paid and cannot be formally and systematically tasked.
CIA: Central Intelligence Agency. US foreign intelligence service. Grew out of the wartime Office of Strategic Services, the OSS, via the Central Intelligence Group, in 1947.
CIS: Commonwealth of Independent States. Formed following the dissolution of the Soviet Union.
COP: Close observation platoon. Non-Special Forces discreet observation capability.
CPS: Covert passive surveillance. Pure intelligence tool deployed by the Int Corps.
DDR: Deutsche Demokratische Republik. Former communist East Germany.
DEFCON: Defense readiness condition. A US strategic alert state matrix from one to five, one being the highest and most serious state.
DIA: Defense Intelligence Agency. US military and defence intelligence agency.

DIS: Defence Intelligence Staff. The UK's military intelligence analysis centre. The US equivalent is the DIA.
DLB: Dead letter box. A short-term hiding place in which to place secret documentation, e.g. micro film. Can be both filled and emptied by an agent and their handler.
DR Congo: The Democratic Republic of the Congo, also referred to as DRC, formerly Zaire, as opposed to the Republic of Congo, also known as Congo Brazzaville.
DS: Directing staff. The teaching staff on military courses.
E1: Int Corps NCOs. E2 were non-Int Corps personnel.
EDA: Emergency deployment area. A pre-recced site to which a unit deploys in times of crisis away from their barracks area.
ERA: Explosive reactive armour. A box containing explosives fitted to vulnerable parts of armoured vehicles that detonates an inbound enemy projectile on contact, thus degrading its effectiveness.
EW: Electronic warfare. Electronic interception of communications at the tactical level, including tactical jamming of an enemy's comms.
F3: Superb Nikon manual camera of its day used by all MOD units. BRIXMIS models sported titanium bodies for enhanced robustness.
F4: The Nikon fully automatic camera that superseded the F3 just prior to the advent of digital film.
FAO: US DIA officer specialisation programme called the Foreign Area Officer. They studied language, history and culture of target countries, becoming area experts. The British did not run a comparable scheme. I am unsure as to the French policy.
FCO: Foreign and Commonwealth Office. The UK equivalent of the US State Department. Based in its beautiful HQ in King Charles Street just off Whitehall, past the Cenotaph.
FKK: *Freikörperkultur.* The German social and health culture dating from the days of the German Empire in the nineteenth century. Belief in nudity and naturism.
FMLM: The French Military Mission. Their equivalent of BRIXMIS.
FRU: Force Research Unit. Latterly an MOD-sponsored unit. Recruitment operation referred to as OP MAXIMISE. Changed its name to Joint Support Group after the Stevens Inquiries.
G1, G2, G3, G4: Army staff branches dealing with admin, intelligence, operations and logistics respectively.

GDR: The German Democratic Republic, the anglicised version of the acronym DDR.
GP: General practitioner. Also known as family doctors. They are the system's gatekeepers with the responsibility to triage more serious medical situations that require the input of specialists.
GRU: Soviet military intelligence. Active across the whole spectrum of intelligence activity including HUMINT and SIGINT.
GSFG/WGF: Group of Soviet Forces in Germany/Western Group of Forces. Soviet military contingent based in DDR until end of the Cold War. Name change took place in 1988.
***Hasi*:** An Albanian clan in the north of the country straddling the border with Kosovo.
HIS: Hostile intelligence service. A foreign enemy intelligence organisation, e.g. KGB.
HQ: Headquarters.
HUMINT: Human intelligence. Generic term for all intelligence derived from interpersonal contact. Sources are exclusively human rather than technical, e.g. IMINT, MASINT, SIGINT.
HVA: *Hauptverwaltung Aufklärung*. The East German foreign intelligence service.
IED: Improvised explosive device. Essentially home-made bombs deployed by terrorist or insurgent groups.
IFOR: The post-Dayton NATO Implementation Force deployed in Bosnia, post-UN.
IGB: Inner German Border. The Cold War-fortified border between DDR and West Germany.
IMINT: Imagery intelligence. Intelligence derived from satellite and aerial recce.
INLA: Irish National Liberation Army.
Int Corps: Intelligence Corps. Never referred to as I Corps. Known as the Green Slime.
IR: Infrared.
IRA: Irish Republican Army. Used as shorthand to denote all non-INLA Irish Republican activity.
JDSC: Junior Division of Staff College.
JIS: Joint Intelligence Staff. JIS (Berlin) took over the role of monitoring WGF from BRIXMIS on its dissolution at the end of December 1990.

JNCO: Junior non-commissioned officer – corporals and lance corporals.
JSIO: Joint Service Interrogation Organisation. Interrogation unit based in Templar Barracks, Ashford.
KaDeWe: Popular name for Berlin's most famous department store, the Kaufhaus des Westens.
KGB: Soviet civilian intelligence service with both a domestic and foreign role. Superseded by SVR, for operations conducted abroad, after the fall of the Soviet Union in 1991.
KZ: *Konzentationslager*. Nazi-era concentration camps first established in 1933. Some continued in use after the end of the Second World War run by the communist authorities.
MAD: Mutual assured destruction. You bomb us and we will destroy you. A crude but effective strategic doctrine that did, however, effectively maintain peace throughout the Cold War.
Manor: The nickname for Specialist Intelligence Wing (SIW), Ashford, derived from the old manor house in which a substantial part of the unit was housed.
MBT: Main battle tank. The T-55, T-62, T-64, T-72, T-80 of WP. Centurion, Chieftain and Challenger of the UK, Abrams of the US, and Merkava of Israeli Defence Forces.
ME: Myalgic encephalomyelitis or chronic fatigue syndrome.
MfS: Ministry of State Security, the East German secret police also known as the *Stasi*.
MILO: Military intelligence liaison officer. Based in DIS with a worldwide trouble-shooting role.
MIO: Military intelligence officer. Army liaison officers working with RUC Special Branch.
MO: Medical officer. The name given to military doctors. Serving MOs attend shortened officer training courses, colloquially (and I suspect historically) known in the Army as the 'Tarts and Vicars Course'.
MOD: Ministry of Defence.
MOD 90: British Military identity card.
NAAFI: Navy, Air Force and Army Institutes. A non-profit-seeking organisation to support the military. In the everyday life of a garrison, the local supermarket. It provided other services including the leasing of cars, which JIS took full advantage of.
NATO: North Atlantic Treaty Organisation. The allied alliance facing off against the Warsaw Pact.

NBC: Nuclear, biological and chemical in the context of conventional warfare. Every Cold War soldier's bogeyman.
NCO: Non-commissioned officer. Military ranks from lance corporal to WO1.
NI: Northern Ireland.
NSDAP: Nationalist Socialist German Workers Party. The Nazi Party.
NVA: *Nationale Volksarmee*. East German army.
OC: Officer commanding. Sub-unit command function supporting the commanding officer.
OOB: Out of bounds.
OP: Observation post. Covert position in urban or rural environment from which to mount static surveillance.
Ops offr: Operations officer. Directs all operational (G3) aspects of a unit.
Ops room: Operations room in which sits the ops offr and any staff he may direct.
Orbat: Order of battle. The formations and units from which a military force is composed.
OTC: Officer Training Corps. University-level version of the Combined Cadet Force. Aims to provide some military exposure to students and generate positive attitudes, if not direct recruitment.
PIRA: Provisional Irish Republican Army. The largest Republican terror group.
POW: Prisoner of war.
PNG: *Persona non grata*. A diplomatic term. You are to be no longer tolerated by your diplomatic hosts. All clauses of diplomatic protection are suspended once declared PNG.
PRA: Permanently restricted area. All the Military Missions operations were restricted by PRA. Encroachment would likely be formally protested.
PX: US military shopping centre. Great prices and great selection of goods from cars to chocolate bars. NATO servicemen enjoy access.
QRF: Quick Reaction Force.
RAF: The Royal Air Force. The 'junior' service, founded in 1918 out of the British Army's Royal Flying Corps and the Royal Naval Air Service.
***Rezidentura*:** The Russian equivalent of a CIA/SIS station. Usually situated within a diplomatic establishment, typically an embassy or consulate, a *rezidentura*, headed by the 'rezident', was the base from

which both GRU and KGB/SVR officers worked. An embassy would invariably support a *rezidentura* from brother services – they were never joint facilities. Defector reports state that the GRU ran five separate *rezidenturas* in West Germany: one each in the embassy in Bonn and the consulate in Köln/Cologne. Significantly, the other three were the Soviet Military Missions, one in each of the three allied former zones of occupation.

RCB: Regular Commissions Board. Located in Westbury, Wiltshire, it selected potential Army officers for training at the Royal Military Academy Sandhurst during its very intensive five-day process. Now called Army Officer Selection Board.

RCT: The Royal Corps of Transport. Soldiers known as 'truckies'. Since 1993, subsumed into the giant Royal Logistic Corps. Soldiers now known as 'loggies'.

RDP: *Régiment de Parachutistes*, a special recce unit of the French Army.

RM: Royal Marines. In my opinion, the most professional infantry unit in the British Military.

RMAS: The Royal Military Academy Sandhurst. The UK's Army officer training equivalent of the US West Point.

RMP: The Royal Military Police. Known fondly as 'monkeys'.

R Sigs: The Royal Corps of Signals. Soldiers known as 'scaleys'.

RTA: Road traffic accident. A vehicle crash.

RUC: Royal Ulster Constabulary. Post-peace accords now the Police Service of Northern Ireland.

RV: Rendezvous.

SAS: Special Air Service. Referred to as 'the Regiment'.

SB: Special Branch. In Northern Ireland, the SB ran all the police agents targeted against both loyalist and Republican terror groups, mirroring the work of the FRU. In the mainland context, Special Branches have always been focussed on Irish terrorism. The first was established within the Metropolitan Police in 1883 as the Special Irish Branch to combat the Irish Republican Brotherhood.

SBA: Sovereign base area. The two UK sovereign territories ceded by the treaty establishing Cypriot independence in 1960.

SED: *Sozialistische Einheitspartei Deutschlands*, The Socialist Unity Party of Germany. The East German Communist party founded in 1949. Post-unification, it tried to change its spots and re-formed as the PDS.

SERB: Soviet External Relations Bureau. The interface between the Soviet Command and the AMLMs.
SF: Special Forces. Formerly comprised SAS, SBS and 14 Coy. Now includes the Special Reconnaissance Regiment (re-roled 14 Coy) and a support group. Directed by Director Special Forces.
SIGINT: Signals intelligence. Refers to all intelligence derived from communications interception.
SITREP: Situation report. Provides updated information.
SIW: Specialist Intelligence Wing. The Army's covert intelligence school based in Templar Barracks, Ashford. Now located in Joint Intelligence Training Group (formerly Defence Intelligence and Security Centre), Chicksands since 1997.
SLR: Self-loading rifle. The Belgian-made 7.62mm British Army rifle replaced by the 5.56mm SA-80.
SMLM-F: Soviet Military Liaison Mission-Frankfurt.
SNCO: Senior NCO. Refers to all ranks from sergeant to WO1.
SO1/SO2/SO3: British Army staff officers at lieutenant colonel, major and captain ranks respectively.
SOE: Special Operations Executive. Second World War intelligence, sabotage and recce organisation.
SOP: Standard operating procedure. An established approach for dealing with an issue.
SOXMIS: Soviet Military Mission. Soviet Military Mission working in British zone of West Germany until end of the Cold War.
SP: Self-propelled in terms of artillery. Guns are mounted on tracked chassis rather than towed.
***Stasi*:** Street name for the East German secret police, formally known as the MfS, the Ministry of State Security.
SUV: Sports utility vehicle.
SVR: The Russian Foreign Intelligence Service. Successor to the KGB 1st Chief Directorate in 1991.
SO2 G2: The officer with the rank of major fulfilling intelligence staff duties in a HQ.
TA: Territorial Army. Vital augmentation for the Regular Army. Referred to unkindly as 'STABs' by their Regular colleagues. Suggested meaning of acronym on a postcard please.

TAOR: Tactical area of responsibility. The area a unit has military operational responsibility for, also called its 'patch'.
TCG: Tasking and Coordination Group. The joint Army/RUC fusion cell directing covert operations in Northern Ireland. There were three in the Province. A great model for combating terrorism.
TRA: Temporary restricted area. Limited-duration restricted areas augmenting PRAs.
UCK/KLA: *Ushtria Clirimtare e Kosoves*, the Kosovo Liberation Army.
UDR: Ulster Defence Regiment. Locally recruited British Army regiment consisting of full-time and part-time members.
UN: United Nations.
US: United States. In UK military parlance, also meaning unserviceable.
USAF: United States Air Force. Not the only US air force; both the US Navy and Marines fly their own aircraft.
USAREUR: The HQ of US forces in Europe, headquartered, since 1952, in Heidelberg until it moved to Wiesbaden in 2012.
USMLM: The US Mission in DDR.
USSR: Union of Soviet Socialist Republics. In Russian Cyrillic, CCCP.
VCP: Vehicle checkpoint. Usually refers to one mounted by police or army.
***VoPo*:** The *Volkspolizei,* the East German civil police. The Missions did not recognise their authority.
VRN: Vehicle registration number. ELECTRIC LIGHT recorded Soviet military VRNs in the Mission. In NI, all vehicles were recorded on the OP VENGEFUL database, accessible to all troops on the ground.
WO1: Warrant officer class one. The most senior non-commissioned rank in the British Army. Incumbents in the British Army referred to as (for example) Mr Jones etc.
WO2: Warrant officer class two. One rank junior to a WO1. Incumbents referred to as (for example) Sergeant Major Jones etc.
WP: Warsaw Pact. The eastern military bloc established in 1955 in response to the formation of NATO.
www.armchairgeneral.co.uk: A great business opportunity in waiting for the discerning entrepreneur and lover of fine furniture. The whole armchair general-related industry is as strong and buoyant as ever, particularly following the challenges of Iraq, Libya, Afghanistan, Syria

and Ukraine. Currently, armchair generals are especially valuing their role and contribution in the war in Gaza/West Bank.

XMG: Crossmaglen. Main town/village in South Armagh, Northern Ireland.

Z-platz: Where BRIXMIS tours slept. Always located discreetly in a forest.

Annex D: Glossary of Equipment

(in order of appearance in the main text)

Ground Equipment

G-Wagen: The Mercedes Geländewagen now manufactured in Austria by Magna Steyr was first launched in 1979. The genesis for the project was apparently the result of a suggestion made by the Shah of Iran, a large Mercedes shareholder. Whatever the history, the definitive 4x4 SUV. Used extensively by the world's military, the Mission versions were distinctly non-standard and were the perfect tool for conveying tourers throughout the DDR.

SS-21 Scarab: The SS-21 was a short-range nuclear-capable ballistic missile entering service in 1975. Maximum range was 70km. The single missile was mounted onto a large, wheeled, amphibious transporter launcher.

Kalashnikov: AK-47 and 74 designed by Mikhail Kalashnikov. The AK-47 entered service in 1949 and is the planet's most widely used rifle. After seventy years, somewhere in the region of 75 million have been produced. Its simple construction makes it a reliable and easy-to-use weapon. Easy to identify in the field, the AK-47 has a distinctive report when fired. The AK-74 began life in 1974 and fires the smaller 5.45mm round. Rifle weighs 3.07kg (compared to 3.47kg of AK-47). Standard magazine holds thirty rounds (twenty on AK-47 due to the larger 7.62mm round). Cyclic rate of fire of 650 rounds per minute. Effective battlefield range of 500m (compared to 350m for the AK-47). Over 5 million produced.

Scud: The Scud is a tactical ballistic missile. The single warhead is launched from a wheeled launcher. The Scud A entered service in 1957, the B in 1964 – BRIXMIS was familiar with this variant. The latest D variant was first deployed in 1989. The B's operational range was out to 300km.
UAZ-469: Soviet jeep omnipresent since 1971. Powered by a four-cylinder 2.45-litre engine. A great all-terrain vehicle. Over 2 million produced. A useful profile and silhouette similarity to the G-Wagen.
Trabant: An iconic East German car. For more information, see the footnote on page 42.
Wartburg: Vehicle company dating back to 1898. It takes its name from Wartburg Castle in Eisenach, deep within the DDR. The name was revived in 1958, and production of saloon cars continued until 1991.
Lada: The brand was originally Russian, starting car production in 1970. The company had a close relationship with Fiat in the early days. The brand survives under Renault's ownership. Its most successful model was the off-road Niva dating from 1977.
T-80: The first gas turbine Soviet main battle tank entering service in 1976. With a weight of 42.5 tonnes, its main armament is a 125mm gun. It has a crew of three and a top road speed of 80km/h. There were 2,256 deployed in the DDR.
Browning: Originally an American company founded in 1878. Taken over by the Belgian Herstal group. Produced the Browning Hi-Power 9mm pistol used by UK forces.
Bedford Transit: Much loved by 'white van men', a full-sized panel van. Direct competitor of the Ford Transit. Bedfords were discontinued in 1987. Used extensively in Northern Ireland as 'covert' troop carriers. Frankly, their intrinsic 'odd' appearance in this context fooled no one.
Humber Box and Snipe: The Heavy Utility Box was the only 4x4 vehicle manufactured by the UK in the Second World War. Adopted by BRIXMIS as its first robust off-roader. The first Snipe was produced for the civilian market in 1938. The Super Snipe also used by BRIXMIS had an impressive top speed of 127km/h. Models continued to be produced until 1967. Humber was part of the Rootes group.
Opel Kapitän/Admiral/Senator: The company was founded in 1862, based in Rüsselheim in West Germany. It was owned by US General Motors from 1929. All capable saloon cars used by BRIXMIS but without the true off-road capability of the G-Wagen. The Senator in particular was heavily modified for Mission service.

Land Rover Discovery/Range Rover: Both 4x4 off-road vehicles but without the robustness and reliability of the G-Wagen.
Ford: Founded in 1903 and based in Dearborn, Michigan. The company pioneered the auto assembly line but their vehicles were no match for German technology in terms of AMLM deployment. USMLM eventually suspended its patriotic approach and followed the other two Missions, buying German.
Chevrolet: Founded in 1911 and part of General Motors. Outselling Ford but equally unsuitable for Mission work.
T-55: The T-55 evolved directly from the T-54. Main armament was a 100mm rifled gun. It had a crew of three, a top speed of 51km/h, and weighed 36 tonnes. The most produced tank in history. Superseded by the T-62.
T-62: A medium tank introduced into service in 1961. The first Soviet tank armed with a smooth-bore gun. Weighed 37 tonnes and had a crew of four. Speed was similar to the T-55. Arguably seen more conflict than any other tank since the Second World War.
BMP-1: The Soviets' first mass-produced tracked infantry fighting vehicle, entering service in 1966. The BMP-1 had a crew of three and could carry eight fully armed infantrymen. Main armament was a 73mm smooth-bore gun. Maximum speed off-road was 45km/h. It was amphibious and NBC proof. Improvements in design resulted in the BMP-2.
BTR-60 PB: First seen in 1961, it was a six-wheeled amphibious APC. Over 25,000 were produced. With a two-man crew, it could carry, like the BMP-1, eight soldiers. Main armament was a 14.5 heavy machine gun. Maximum speed was 80km/h. Used extensively in Afghanistan.
2S1: Self-propelled amphibious artillery piece on a tracked MT-LB chassis, first in service in 1972. Gun was a 122mm howitzer, based on the towed, D-30, with a range out to 21.9km. Weighed 18 tonnes with a top speed of 30km/h off-road. Extensive campaign service in the Balkan Wars and the Middle East.
2S-3: The 2S-3 entered service one year before the 2S-1. It was also a tracked SP gun, of 152.4mm calibre and with a range of 18.5km. It was heavier at 28 tonnes. Off-road speed was 45km/h. Crew was four. Similar war record to its smaller brother, the 2S-1.
MT-LB: Introduced in the 1950s, the MT-LB was a workhorse, multipurpose, fully amphibious, armoured tracked vehicle. It had a crew of two and could carry eleven troops in the rear. Mounted a single

machine gun or cannon. Weighed 11.9 tonnes and could reach speeds of 61km/h off-road.
GP-5 Schlem: A horribly uncomfortable but effective early-generation NBC respirator that featured a hose leading to the external filtration canister.
BMP-2: Successor to the BMP-1, and redesigned based on experience in combat during the Yom Kippur War. Introduced in the early 1980s. Weight of 14.3 tonnes and top speed of 65km/h on road and 45km/h off-road. Able to swim at a steady 7km/h. Crew of three, with seven troops in the rear compartment. Main armament a 30mm autocannon. Another war-proven Soviet vehicle.
Lada Niva: Dating from 1977, the Niva was the world's first mass-produced off-road civilian vehicle. Produced with three petrol engine variants, up to 1.8 litres and a 1.9 diesel.
JS-2: A heavy tank at 46 tonnes, designed for breakthrough operations. It fired a 122mm main gun. Deployed extensively at the end of the Second World War in the advance on Berlin. Crew of four and a top speed of 37km/h.
T-54: See T-55.
PMP: A lorry-mounted pontoon bridge introduced into service just after the end of the Second World War. Maximum weight-carrying capacity was 60 tonnes.
SPG-9: A 73mm tripod-mounted recoilless gun introduced in 1962. Man-portable, it fired fin-stabilised HE or high-explosive anti-tank shells. Total weight, with tripod, of 59.5kg and effective range of 800m.
Frog-7: Short-range unguided rocket system mounted on an 8x8 lorry. In service from 1964, the Frog was designed for nuclear delivery but was modified as a conventional weapon. Maximum firing range was out to 65km.
BM-22: A 220mm multiple rocket launcher, lorry mounted, superseding the BM-21, it entered service in 1975. It packed sixteen rockets with a range of 35km. Salvo rate was twenty seconds.
BRDM-2: A wheeled four-man amphibious armoured scout car on the scene since 1962. Mounted a 14.5 heavy machine gun. Top speed of 100km/h. Not a particularly effective battlefield asset, but was used to tow Brezhnev's coffin.
AT-2 Swatter: The AT-2 Swatter semi-automatic command to line of sight (SACLOS) radio command, solid fuel, anti-tank missile, designed for

deployment on helicopters, came into service in 1964. With a weight of 27kg, it travelled at 160m/s at ranges from 500m to 2.5km.

T-72: A widely exported MBT that entered service in 1971. Main armament was a 125mm smooth-bore gun. Weighed 41.5 tonnes with a crew of three. The gun was served by an autoloader. A low silhouette encouraged the myth that crew members could be no taller than 5ft 3in (160cm). Soviet figures state that crew members could be as tall as 5ft 9in (175cm) and still operate effectively. The T-72's maximum speed was 60km/h. A total of approximately 25,000 tanks were built. Only NVA deployed T-72s in the DDR.

T-64: Second-generation MBT introduced to Soviet tank divisions in 1964. Armed with a 125mm smooth-bore gun and its revolutionary autoloader firing eight rounds per minute, it had a crew of three. The T-64B, which came into service in 1976, could fire anti-tank guided missiles through its main gun. It employed composite armour. They were expensive tanks with an advanced power pack – the T-72 was 40 per cent cheaper. At 38 tonnes it was relatively light and could move at 60km/h depending on the version. The T-64 was never exported and remained deployed solely by Soviet forces. The T-80 was a later derivative.

BTR-70: An eight-wheeled APC superficially similar to the BTR-60 in appearance, entered service in 1972. Crew of three and seven troops in the main compartment. Main armament was a 14.5mm machine gun. Weight of 11.5 tonnes and top speed of 80km/h. The BTR-70's poor performance in Afghanistan was the catalyst for the BTR-80, which was never deployed into the DDR.

2S-5: A turret-less 152mm SP gun built onto a modified SA-4 Krug chassis with a longer range and higher rate of fire than the 2S-3. Effective range of 28km and a fire rate of five or six rounds per minute. It was first deployed in 1978. Weight 28.2 tonnes with a speed of 25km/h off-road. Five-man crew with an additional two in an ammunition carrier. The SP version of the M 1976 152mm.

RKhM: A tracked chemical, biological, radiological and nuclear recce vehicle.

AT-7: A man-portable, wire-guided, tube-launched SACLOS 94mm anti-tank weapon in service since 1979. Can be shoulder-fired or mounted on its tripod. Effective battlefield range from 40m out to 1,000m. Not considered an effective weapon system due to poor range and accuracy. Unfavourably compared to the NATO Milan system.

Milan: A joint French–German project that came into service in 1972. A wire-guided SACLOS anti-tank weapon. The operator was required to have direct sight of his target. Battlefield range was 200m to 2,000m.
SS-21: A wheeled tactical ballistic missile dating from 1976. Warheads could deliver chemical, nuclear or fragmentation loads. Range was out to 70km. First deployed in DDR in 1981.
M 1976 152mm: Also known as the 2A36 Giansint, this towed artillery gun had a distinctive double-wheel chassis and split trail. In service in 1975. Calibre was 152.4mm, effective out to 40km. SP version was the 2S-5. Rated as an effective weapon system.
PMM-2: A floating bridge system or individual single-vehicle ferry built onto a PTS-2 tractor. Up to three units could be joined together to simultaneously ferry a total weight of 127.5 tonnes (essentially three T-80s).
K-611: A radiological recce vehicle based on the MT-LB chassis.
2S-9: A lightweight SP amphibious air-droppable 120mm gun-mortar, with a range of 8.8km, entering service in 1981. Had a crew of four. Weight 8.7 tonnes. Top speed 60km/h. The unconventional hybrid mortar/howitzer had no NATO equivalent.
MTLBu: An updated more powerful version of the MT-LB chassis. It is the chassis for a phenomenal variety of equipments, from command vehicles to specialist artillery support vehicles et al. Top speed 61.5km/h, and 30km/h off-road. During the BRIXMIS course arguably one of the trickiest customers in terms of learning all the diverse variants.
PRP-3: A forward observation vehicle whose role was to identify forward enemy positions and the location of artillery units for counter-battery fire. Based on the BMP-1. A crew of five with a larger turret than the BMP-1 to accommodate its observation devices and ground surveillance radar. Armed with a 7.62mm machine gun for self-defence in its exposed forward positions. Entered service in 1972. Succeeded by the PRP-4.
SS-23: Brought into service as a truck-mounted, theatre, nuclear-capable, short-range ballistic missile to replace the Scud-B in 1987. It was a potentially very effective weapon, accurate and virtually impossible to intercept. Range was 500km.
TM-62: An anti-tank mine packing 7.5kg of explosive. Mines were laid manually or by automated minelayer or helicopter.
RPG-7: The ubiquitous shoulder-fired 40mm anti-tank rocket-propelled grenade launcher that first came onto the scene in 1961. Weight is 7kg.

Effective from 10m out to 700m. In excess of 9 million were produced. An updated version (designated the RPG-7USA) was unveiled by US company Airtronic USA in 2009.
GAZ trucks: Soviet truck company founded in 1932. The 4x4 GAZ-66 light truck was seen everywhere in the DDR in both military and civilian use. Since 1964 over 1 million were produced.

Air Equipment

Sputnik: The first artificial Earth satellite launched by the Soviets in 1957, marking the start of the space race. Weighing 83.6kg, it orbited for three weeks before its batteries ran down.
Apollo: Project Apollo was the third US-manned space programme succeeding Mercury and Gemini. Lasting from 1961 to 1972, it cost an estimated $156 billion but succeeded in putting man on the moon six times.
MiG: Abbreviation of the Mikoyan aircraft company. Founded in 1939, the company has produced some of the most iconic military aircraft including the MiG-15 and MiG-29.
B-52: Known as the Stratofortress, Boeing's strategic bomber is expected to still be in service 100 years after first flying in 1952. It can carry a 32,000kg payload and operate out to 14,000km without refuelling. A key Cold War warrior, Soviet air defence advances saw the B-52 re-roling as a low-level penetration bomber.
U-2: The Lockheed U-2 is an all-weather high-altitude reconnaissance aircraft cruising up to 70,000ft. No more than 200 pilots have operated the aircraft since its inaugural flight in 1955. Early flights were flown by CIA contract pilots but now the aircraft is a USAF asset. Two aircraft have been shot down, both during the Cold War (Powers and Anderson). The newest variant was upgraded in 2012. A total of 105 U-2s have been built.
Phantom: Entering service in 1961, the McDonnell Douglas F-4 is a two-seat, twin-engine, all-weather, long-range supersonic interceptor and fighter-bomber with a top speed in excess of Mach 2.2. With over 5,000 built, it is the most produced US supersonic plane in history. The Wild Weasel variant was operational until 1996. The Phantom saw service in all three US air forces (USAF, Marines and Navy), and was widely purchased

by foreign nations including the UK, Israel, Germany, Turkey, South Korea, Japan, Greece and the Shah's Iran.

Victor: The Handley Page Victor flew from 1953 until 1993. Initially deployed as a strategic bomber as part of the UK's 'V Force', the airborne nuclear deterrent, alongside the Valiant and Vulcan. When the Royal Navy assumed the nation's nuclear deterrent with Polaris, the V Force was retired, and some Victors were re-roled as aerial refuelling tankers until this role was taken on by a combination of VC10s and Tristars.

Vulcan: The Avro Vulcan was also a member of the UK's V Force. With its highly distinctive delta wing, it was operational from 1956 to 1984. After Polaris assumed the nation's primary nuclear role, Vulcans continued to have a tactical nuclear role in support of NATO ground forces. It could carry twenty-one 1,000lb bombs. One example was kept in flying condition after its operational retirement, and was an amazing sight to see in flight at air shows, with its bomb doors open making slow-speed turns. The Vulcan is famous for its attack on Port Stanley during the Falklands War. A total of 136 were built.

VC10: Retiring from service in 2013, the Vickers VC10 was flown by the RAF for fifty-one years. It was a long-range, mid-sized, narrow-body jet airliner, with a distinctive rear-engine quad arrangement. Until the development of Concorde, the VC10 was the world's fastest airliner. A number of civilian airlines operated the plane, including BOAC. The RAF had a fleet of fourteen. It was a unique experience to fly in, with seats facing backwards for safety reasons. Before retirement, VC10s were used as aerial tankers and during the Iraq conflict as medevac aircraft.

Concorde: The world's first commercial supersonic passenger jet. Operated by British Airways and Air France, it carried passengers from 1976 to 2003. With a maximum speed of over Mach 2, a truly iconic aircraft. Under the treaty, twenty (including six prototypes) were built in a cooperative venture between the British Aircraft Corporation and France's Aérospatiale. With huge delays and problems in development, the programme cost in excess of £310 million, the figure adjusted for inflation. It is ironic that the fatal crashes that ended the life of the Soviet Tu-144 'Concordski' in 1973 and effectively Concorde in 2000 were both in Paris.

Apache: The Boeing AH-64 Apache is a US-made twin-turboshaft attack helicopter. It has a crew of two in a tandem cockpit. The Apache entered service in 1986 and has since accumulated an impressive operational record.

Main armament is a forward-firing 30mm chain gun. It has four hardpoints for mounting a mix of Hellfire or Hydra 70 rockets. The Apache was developed by Hughes to replace its AH-1 Cobra. It is operated by a number of other nations including the UK, which builds the helicopter under licence at Westland. The AH-64D, considerably more effective than the A version, is currently being replaced by an E variant. One revolutionary fighting feature of the AH-64D was the integrated helmet and display sighting system slaving gun targeting to the positioning of the gunner's head. Maximum speed was 293km/h with a combat range of 480km.

Wessex: The Westland Wessex was a Sikorsky H-34 built under licence in the UK. Deployed from 1961 to 2003, it saw service in Borneo, then the Falklands and extensively in the border areas of Northern Ireland with its ability to carry sixteen troops. Maximum speed was 212km/h.

Gazelle: A French five-seat light utility/recce helicopter that can be fitted with anti-tank missiles produced by Aérospatiale. It will have been deployed by UK forces from 1973 until its planned retirement in 2025. Maximum speed is 310km/h at sea level.

Scout: The Westland Scout was a light utility helicopter deployed from 1963 to 1994. It formed the backbone of the Army Air Corps, seeing service in all the UK's 'bush fire' wars. In Northern Ireland it was used to mount the aerial Heli-Tele surveillance system and the 3.5 million candlepower Nightsun searchlight. The Scout could carry four soldiers crammed in behind the crew and had a top speed of 211km/h.

Chipmunk: The de Havilland Canada DHC-1 Chipmunk is a tandem, two-seat, single-engine primary trainer. It was designed to supersede the Tiger Moth biplane. The UK used the aircraft from 1946 until 1996. The Chipmunk's top speed was 166km/h, even slower than the average helicopter. It proved to be a key tool in the BRIXMIS box.

Mil Mi-24 Hind: A large attack helicopter gunship capable of carrying eight troops and known to Soviet pilots as the 'flying tank' or 'Crocodile'. It entered service in 1972. Mikhail Mil realised the solution to the requirement for increased battlefield mobility was a flying infantry fighting vehicle. He faced opposition from conventional traditionalists within the Soviet military but his concept prevailed. The Hind-D was easily recognisable by its double bubble cockpit canopy. The Hind was both fast, with a top speed of 335km/h, and well armoured. Most variants were armed with a Gatling gun plus an additional autocannon. The D mounted two 12.7 machine guns in addition to the Gatling, plus a pair of S-5 rockets and nine M17 Fleyta on each wingtip pylon. Arguably the most combat-seasoned attack helicopter in the world, the Hind was a pivotal weapon system in the Soviet Afghan invasion with up to 250 deployed in theatre.

Sukhoi Su-24 Fencer: The first Soviet day or night, all-weather supersonic attack aircraft with a variable wing. The crew of two sat side by side. It came into service in 1974. The later E variant was a recce version of the aircraft and the F an ELINT variant. The aircraft was very similar to the General Dynamics F-111. Maximum speed was 1,654km/h at high altitude. The Fencer was protected by a 23mm rotary cannon. It had hard points for rockets, missiles and bombs.

General Dynamics F-111: No longer in service, the F-111 was a supersonic, medium-range interdictor and tactical attack aircraft that was also deployed as a nuclear bomber, with recce and electronic-warfare variants. The USAF first flew the plane in 1967. The F-111 used variable-sweep wings, afterburning turbofan engines and automated terrain-following radar for high-speed, low-level sorties, paving the way for next-generation aircraft. Its internal bomb bay could carry 2,750lb conventional bombs, or a single nuclear weapon. It had a removable 20mm cannon and could carry two Phoenix long-range air-to-air missiles for self-defence. The F-111 played a significant role in the war in Indo-China and deployed during the first Gulf War. The Australian Air Force also operated the plane until finally replacing it with the F-18 in 2010. The F-111 had a maximum speed at high altitude of Mach 2.5.

C-130: The Lockheed C-130, also commonly known as the Hercules, is one of the great success stories of aviation history. Since its development there have been over forty variants, including use as a gunship, and extensive civilian usage. The aircraft was designed as a military troop and cargo transport capable of using unprepared runways. The Hercules came into service in 1956 and over 2,500 airframes have been manufactured. It is the longest continuously produced military aircraft in history. In the 1990s the J version was developed with new turboprop engines, six-bladed propellers and enhanced avionics. The J was used by the RAF's Special Forces flight until 2023. The four-engine C-130 can lift 19,000kg, ninety-two 'light' passengers or sixty-four airborne troops. Cruising speed is 541km/h with a range of 3,800km.

Yak-28 Brewer: A swept-wing turbojet employed initially as a medium tactical bomber. Later versions included recce, EW and interceptors.

Introduced into service in 1960, it had a crew of two and a maximum speed of 1,840km/h. Fitted with up to four air-to-air missiles.
Sukhoi Su-17 Fitter: The Fitter was the first variable-sweep wing fighter-bomber, which entered Soviet service in the 1970s. Variants included the B, C, E (a two-seat trainer) and K (upgraded avionics). It was capable of dropping a free-fall nuclear bomb and additionally had a 'toss bomb' capability. Armed with four autocannon, it had external hardpoints for 4,000kg of bombs or rockets. Maximum speed was 1,860km/h at altitude.
MiG-21 Fishbed: The iconic MiG-21 was a supersonic jet fighter and interceptor with a top speed of 2,175km/h. It entered service in 1959. Somewhere in the region of sixty nations operated the aircraft, and it is still operational in some. It was armed with an array of rockets and missiles and one 23mm internal autocannon.
An-26 Curl: The Curl was a twin-engine, high-wing transport aircraft. It became operational in 1970. There were numerous variants for both civil and military use. It could be fitted with external bomb racks. Flown by a crew of five, it could carry forty passengers or 5,500kg of cargo with a range of 2,500km. Cruise speed was 440km/h.
DR-3: No open source detail.
Mil Mi-8 Hip: Amongst the most produced helicopters in the world since its introduction into service in 1967, serving over fifty countries. The Hip was designed as a utility helicopter but variants fly as airborne command posts, gunships and recce platforms. It was the first rotary design to have purpose-designed engines rather than 'borrowed' aero engines. A significant factor in the approval for the new design was the success of the Bell UH-1 as a battlefield taxi in the Vietnam War. The Hip has a crew of three and can carry twenty-four passengers or 4,000kg internally and on external hardpoints. Top speed is 240km/h with a range of 495km.
MiG-23 Flogger: A variable-geometry fighter, the first in the Soviet inventory to be fitted with a look-down/shoot-down radar and one of the first to be armed with beyond-visual-range missiles. It first flew with the Soviet Air Force in 1970. It was commissioned to counter the MiG-21's primitive radar, short range and limited weapons load. Maximum speed was 2,499km/h. The Flogger was armed with one 23mm autocannon and six hardpoints for rockets or missiles, or for a single 500kg bomb per hardpoint.
MiG-25 Foxbat: One of the fastest aircraft in the world at Mach 2.83, it served with the Soviet Air Force as a supersonic interceptor and recce

aircraft from 1970. The Foxbat had a powerful radar and was armed with four air-to-air missiles. NATO overestimated the aircraft's capability until a defecting Soviet pilot kindly brought one to the West. The Foxbat was designed to deter US high-altitude overflights over Soviet territory, hence its operational ceiling of 89,000ft. Its high performance brought enormous propaganda value to the USSR – the Foxbat held, in all, twenty-nine aviation records, several of which remain un-bettered to this day.

Sukhoi Su-25 Frogfoot: The Frogfoot was a subsonic, single-seat, close-air-support, twin-engine aircraft that entered service in 1981. It saw extensive service in Afghanistan. Previous close-support aircraft were regarded as inferior as they were poorly protected from ground fire and were too fast for accurate mission execution. Heavily armed for its role, the Frogfoot carried a 30mm autocannon, two 23mm machine guns with eleven hardpoints to mount a total of 4,400kg of ordnance combining rockets, air-to-air, air-to-surface and anti-radiation missiles, and bombs. It also carried chaff-and-flare dispensers to enhance protection.

MiG-29 Fulcrum: A twin-engine air superiority fighter designed to counter the US F-15 and F-16 when it came into service in 1977. It quickly broadened its role to become a multi-role fighter. After the fall of the Wall, former East German Fulcrums continued flying with the German Luftwaffe. They performed well against US fighters dogfighting on exercise. Maximum speed is 2,400km/h. Armaments consist of a 30mm autocannon and seven hardpoints for rockets, missiles and bombs (6 x 665kg).

F-15: Built by McDonnell Douglas, the F-15, with its distinctive twin tail, is an all-weather tactical fighter. It has served with US forces since 1976. It is the most successful modern fighter with over 100 kills and no losses in combat, thanks largely to the record of the Israeli Air Force. It can fly at high altitude at Mach 2.5. It is armed with a six-barrel rotary cannon and has nine hardpoints for other ordnance up to 7,600kg.

F-16: A multi-role fighter built by General Dynamics, although it started life in 1978 as an air superiority fighter utilising the exceptional field of view from its bubble canopy. It is the first fighter to be intentionally designed to be aerodynamically unstable to improve manoeuvrability. It can achieve speeds of Mach 2.05. Armament potential is similar to the F-15.

Beriev A-50 Mainstay: As an AWACS aircraft, the Mainstay was based on the Il-76 transport. It entered service in 1984. A total of approximately forty were built. The fifteen-man crew process the data from the 9m

over-fuselage rotodome, controlling up to ten fighter aircraft. It has a range of 7,500km.
Tornado: The tri-nation,variable-sweep, multi-role combat aircraft developed by the UK, Germany and Italy. The Panavia Tornado was designed as a fighter-bomber, with a crew of two, to suppress enemy air defences, as a recce platform and to intercept enemy aircraft. It came into service in 1979. Since then approximately 1,000 were built. At an altitude of 30,000ft, maximum speed is 2,400km/h. In terms of armament, the Tornado was cleared to carry just about every type of air-launched weapon in the NATO armoury.
BAe125: A twin-engine, mid-size business jet originally developed by de Havilland, also known as the Dominie. Standard cabin layout accommodated eight passengers. Cruising speed is 841km/h. The RAF's 32 (The Royal) Squadron operated a number.

Air Defence, Radar and Command and Control

Snow Drift: Supporting the SA-11 Gadfly SAM system, Snow Drift is a mobile 3D radar using an electronically steered pencil beam. It has the capability to track up to fifty targets simultaneously, providing target designations on six of those targets. It can see out to 85km and detect low-flying air targets at 100m at 35km.
SA-10 Grumble: The SA-10 Grumble is a long-range surface-to-air missile also known as the S-300. It is capable of engaging a number of targets at low to medium altitude. The SA-10 can target and hit low radar cross-section targets such as cruise missiles and tactical ballistic missiles like the US Pershing, potentially with the capability to hit some types of strategic ballistic missile. The missile is launched vertically and driven by a single-stage solid motor. The warhead carries 100kg of high explosive (HE). The missile can hit targets within a 25m to 30,000m height envelope out to at least 90km. The mobile battery is operational within five minutes of halting. First deployed in the USSR in 1980. Still regarded as a first-rate weapon.
Flap Lid: The multi-function phased array radar with digital beam supporting the mobile SA-10 Grumble surface-to-air missile system. Flap Lid B was mounted on a large ten-wheeled truck, lying flat when travelling. It is capable of tracking six targets out to 300km, against

each of which two missiles can be fired. The radar is highly resistant to electronic countermeasures.

SA-11 Gadfly: Designed as a self-propelled, medium-range, surface-to-air weapon, replacing the SA-6 Gainful, to target cruise missiles, smart bombs, fixed- and rotary-winged aircraft and drones. Up to three missiles could be guided onto a single target. Each mobile launcher consisted of four missiles, and was crewed by four operators. Aircraft could be targeted within an engagement zone of 15m up to 25,000m, and out to 42km. Introduced into service in 1980.

SA-2 Guideline: A high-altitude air defence system capable of hitting targets flying up to 82,000ft (25,000m), first deployed in 1957. Initially developed to counter US strategic bombers, which were able to strike targets deep within the Soviet Union. It became the most widely deployed air defence weapon in the world. Warhead weight was 200kg. Famous in Cold War mythology for downing Gary Power's U-2, and that piloted by Rudolf Anderson overflying Cuba in October 1962 – Anderson was killed.

Long Track: A tracked mobile early warning and target acquisition radar introduced in 1963. It was the first high-mobility radar deployed by Soviet Forces. Supported the SA-4.

Thin Skin B: A truck-mounted, height-finding radar introduced in 1970 with a range out to 296km. Supported the SA-4.

SA-4 Ganef: The SA-4 was a tracked, mobile, medium-range, out to 55km, medium- to high-altitude SAM system. The missile could hit targets up to 80,400ft. Each chassis carried two missiles. Warhead weight was 135kg. The Ganef entered service in 1965.

SA-6 Gainful: A battlefield self-propelled low- to medium-level SAM designed to defend ground forces from air attack. Effective range was 24km up to 46,000ft. Each launcher mounted three missiles. Gainful entered service in 1967.

SA-9 Gaskin: A short-range, highly mobile, low-altitude SAM. Gaskin was mounted on a wheeled BRDM-2 amphibious vehicle. Missiles were launched from a pair of launchers, each holding two warheads. Entered service in 1968. Engagement zone was 900m up to 4,200m and out to 6,500m.

Twin Ear: No open source detail.

Turn series: No open source detail.

Wood Bine: No open source detail.

SA-13 Gopher: Another highly mobile, short-range SAM system. Mounted on a MT-LB tracked chassis, designed primarily to defeat low-altitude threats such as helicopters. Each launcher mounted four 5kg warhead missiles. An additional eight warheads were carried inside the vehicle. Missile was effective out to 5km up to 11,500ft. In service from 1976.
Fan Song: Distinctive, double-domed, trailer-mounted fire control and tracking radar supporting the SA-2. Capable of guiding up to three missiles simultaneously.
Tin Shield: A semi-mobile radar supporting SA-10, designed to detect targets at low, medium and high altitude out to 350km.
SA-5 Gammon: The Gammon was a very long-range, out to 300km, medium- to high-altitude SAM system commissioned to defend large-area strategic targets, such as cities or industrial or military installations, from bomber attack. It was capable of hitting any aircraft within an envelope extending up to 130,000ft. Each SA-5 equipment was a fixed single-rail launcher. The warhead weighed a hefty 217kg. The SA-5 came into service in 1966. Interestingly, the Iranians claim to have developed a mobile launcher for the system.
2S-6: The 2S-6 is a hybrid, self-propelled, anti-aircraft weapon with a crew of four, armed with both air defence missiles and guns. It was designed to provide organic day and night, all-weather cover for infantry and tank units against low-flying aircraft, helicopters and cruise missiles. The 2S-6 is armed with eight missiles, with a range of 8km up to 11,500ft, and two 30mm guns. The guns can be fired when the system is mobile; however, to deploy its missiles, it must be stationary. It is capable of attacking six targets simultaneously. The 2S-6 entered service in 1982.
Silver Box: No open source detail.
Copper Log: No open source detail.

Misc

Polaris: The UGM-27, two-stage, solid-fuelled, nuclear-armed, submarine-launched ballistic missile manufactured by the Lockheed corporation and deployed by both the US and Royal British Navy. Just under 10m long, the Polaris could deliver 3 x 200kt nuclear devices to targets 4,600km away

with an accuracy of 910m. Missiles could be launched from a submerged and moving submarine. The programme began development in 1956, with the first missiles in service in 1963. Polaris was replaced in the US by Poseidon from 1972, then subsequently by Trident with its much larger range and size. The UK bought the missile but manufactured its own warheads for its four nuclear submarines. The UK replaced Polaris directly with Trident.
Cruise: A generic term for guided missiles used against terrestrial targets that are designed to deliver a large warhead, accurately over long distances. Cruise weapons fly at supersonic, or very high subsonic, speeds and are capable of maintaining low-level, low-trajectory flight. A considerable number of countries have developed and deploy cruise technology.
Leica: The company has made top-quality cameras, binoculars, lenses, rifle scopes and microscopes since 1869. It is based in Wetzlar in Germany. Prior to 1986 the company was called Leitz.
Modulux: Bulky image intensifier casting a green glow onto images. The Nikon F3 could be fitted directly to the rear of the tube, enabling night-time photography. The Mini-Modulux was smaller and lighter. Now Stone Age technology in comparison to new-generation optics.
Zeiss: Founded in Jena in 1846. Current headquarters is in Oberkochen near Stuttgart. Still has a substantial presence in Jena in post-reunification eastern Germany. The company began making microscopes but by the First World War was the world's largest camera manufacturer. Still famous for the high-quality binoculars and cameras and other lenses that it produces.
F3: The Nikon F3 was the third professional 35mm single-lens reflex TTL camera body, introduced by the company in 1980. It had manual and semi-automatic aperture priority control. BRIXMIS used the F3/T made of titanium. The MD-4 external motor drive unit was fitted to provide continuous shooting.
F4: The Nikon F4 was the successor to the F3 in the company's professional range. It was introduced in 1988. It was the first Nikon to incorporate a practical autofocus system. Nikon did make a digital version for NASA for use on the Space Shuttle.

Index

The History Press
The destination for history
www.thehistorypress.co.uk